Sustainable Development Goals Series

The **Sustainable Development Goals Series** is Springer Nature's inaugural cross-imprint book series that addresses and supports the United Nations' seventeen Sustainable Development Goals. The series fosters comprehensive research focused on these global targets and endeavours to address some of society's greatest grand challenges. The SDGs are inherently multidisciplinary, and they bring people working across different fields together and working towards a common goal. In this spirit, the Sustainable Development Goals series is the first at Springer Nature to publish books under both the Springer and Palgrave Macmillan imprints, bringing the strengths of our imprints together.

The Sustainable Development Goals Series is organized into eighteen subseries: one subseries based around each of the seventeen respective Sustainable Development Goals, and an eighteenth subseries, "Connecting the Goals", which serves as a home for volumes addressing multiple goals or studying the SDGs as a whole. Each subseries is guided by an expert Subseries Advisor with years or decades of experience studying and addressing core components of their respective Goal.

The SDG Series has a remit as broad as the SDGs themselves, and contributions are welcome from scientists, academics, policymakers, and researchers working in fields related to any of the seventeen goals. If you are interested in contributing a monograph or curated volume to the series, please contact the Publishers: Zachary Romano [Springer; zachary.romano@springer.com] and Rachael Ballard [Palgrave Macmillan; rachael.ballard@palgrave.com].

Edward Debrah Wiafe

Wildlife Conservation and Agriculture Development in Africa

Springer

Edward Debrah Wiafe
School of Natural and Environmental Sciences
University of Environment and Sustainable Development
Somanya, Ghana

ISSN 2523-3084 ISSN 2523-3092 (electronic)
Sustainable Development Goals Series
ISBN 978-3-032-09232-8 ISBN 978-3-032-09233-5 (eBook)
https://doi.org/10.1007/978-3-032-09233-5

This Springer imprint is published by the registered company Springer Nature Switzerland AG
The registered company address is: Gewerbestrasse 11, 6330 Cham, Switzerland

Dedication

To my family

Foreword

Africa's natural wealth is unrivaled—from its expansive savannas to its intricate tropical rainforests, and from iconic megafauna to lesser-known endemic species. Yet, the continent faces a dual challenge: conserving its extraordinary wildlife while ensuring agricultural productivity and food security for its growing population.

In this compelling volume, "Wildlife Conservation and Agricultural Development in Africa," Professor Edward Debrah Wiafe presents a timely, well-researched, and context-specific discourse on the coexistence of biodiversity conservation and sustainable agriculture.

This book not only maps out Africa's ecological treasures and threats but also provides pragmatic insights into how policy, traditional knowledge, community-based practices, and innovation can harmoniously drive both conservation and development.

As the Vice-Chancellor of the University of Environment and Sustainable Development, I commend this work for its academic rigor, accessibility, and relevance to Africa's urgent developmental and environmental priorities. It is a vital resource for researchers, students, policymakers, and practitioners committed to a sustainable and biodiverse future. I am confident that this book will serve as a beacon for informed discourse, inspire transformative action, and catalyze integrated solutions across the continent.

University of Environment and Sustainable Development
Somanya, Ghana

Prof. Eric Nyarko-Sampson

Preface

Africa stands at a unique crossroads—a continent rich in biodiversity yet equally challenged by the urgent demands of agricultural expansion and food security. In the face of accelerating environmental change, growing human populations, and intensifying pressures on land and resources, the need for innovative strategies that harmonize wildlife conservation and agricultural development has never been more critical.

This book, *Wildlife Conservation and Agricultural Development in Africa*, is born from the recognition that sustainable development and conservation are not mutually exclusive goals, but interdependent pillars for a resilient future. Across its chapters, this book explores the deep-rooted biodiversity riches of Africa, the complexities of human–wildlife conflict, the evolving agricultural landscape, and the intricate balance required to achieve both food security and ecosystem integrity.

Drawing upon case studies, policy analyses, traditional knowledge, and contemporary innovations, this volume offers a comprehensive examination of how Africa can forge pathways toward sustainable farming practices that support both human livelihoods and the continent's irreplaceable wildlife heritage. Special attention is given to the role of indigenous knowledge systems, community-based conservation initiatives, and the challenges and opportunities presented by climate change.

This book is intended for a diverse readership: researchers, policy-makers, conservationists, agricultural practitioners, students, and anyone passionate about Africa's future. It invites readers to reimagine the relationship between people, agriculture, and nature—not as conflicting forces, but as potential partners in shaping a more just, food-secure, and biodiverse Africa.

It is my hope that the insights contained within these pages will inspire new conversations, innovative actions, and enduring commitments to both conserving Africa's extraordinary wildlife and achieving sustainable agricultural development.

Somanya, Ghana — Prof. Edward D. Wiafe

Acknowledgements

The journey to completing *Wildlife Conservation and Agricultural Development in Africa* has been marked by generous support, encouragement, and collaboration from many individuals and institutions across the continent and beyond.

I extend my deepest gratitude to the University of Environment and Sustainable Development (UESD) for providing the academic environment, institutional resources, and intellectual freedom that made this work possible. I also appreciate the collegiality and critical input of faculty and research associates whose insights helped shape many chapters of this book. Special thanks are due to community members, local farmers, and conservation officers across Ghana and other African regions whose lived experiences, stories, and practical knowledge brought authenticity and nuance to the discourse. Their voices have been invaluable in linking theory to real-world challenges and opportunities in the field. I wish to thank my colleagues in biodiversity conservation, agroecology, and land management, particularly those who contributed through collaborative research, fieldwork, and peer review. Their shared commitment to sustainable development in Africa served as a constant source of motivation. I am especially grateful to development partners, environmental NGOs, and international researchers who have been long-standing allies in promoting harmony between conservation and agricultural advancement. Your ongoing work continues to inspire and enrich the conversation around integrated land use and ecological stewardship. To my family, who continue to support me with patience and understanding during long periods of research and writing, I am forever thankful. Finally, this book is dedicated to the young African scientists, conservationists, and agricultural innovators whose passion and vision for a balanced, biodiverse, and food-secure continent will shape Africa's sustainable future.

Contents

List of Abbreviations

AGRA	Alliance for a Green Revolution in Africa
CBD	Convention on Biological Diversity
CITES	Convention on International Trade in Endangered Species of Wild Fauna and Flora
DRC	Democratic Republic of the Congo
FAO	Food and Agriculture Organization of the United Nations
GDP	Gross Domestic Product
GIS	Geographic information system
HWC	Human–wildlife conflict
IUCN	International Union for Conservation of Nature
NGO	Non-governmental organization
PA	Protected area
R&D	Research and Development
SADC	Southern African Development Community
SDG	Sustainable Development Goals
UN	United Nations
UNDP	United Nations Development Programme
UNESCO	United Nations Educational, Scientific and Cultural Organization
USAID	United States Agency for International Development
WCS	Wildlife Conservation Society
WTO	World Trade Organization
WWF	World Wide Fund for Nature

List of Figures

List of Tables

The Biodiversity Riches of Africa

1

1.1 Introduction

Africa is renowned for its rich and diverse wildlife, making it a popular tourist destination and appearing on many people's bucket lists of places to visit in their lifetime. The continent is home to some of the most iconic wildlife in the world, including the African elephant, the lion, and the mountain gorilla. However, Africa is not just about big game, there are also a huge variety of unique and wonderful plants, reptiles, birds, and insects that call Africa home, many of which are found nowhere else on Earth (Bedair 2023). Biodiversity is the term used to describe the variety of life found on Earth, at all its different levels, from ecosystems to species. And Africa is a continent that has this in abundance! This diversity is due to the varied and unique climatic, geographical, and geological factors that have played important roles in shaping the continent's diverse ecosystems. Although Africa is known for its rich biodiversity, it is also faced with a variety of threats, including climate change, pollution, infrastructure development, and land misuse (Archer et al. 2018). This chapter—"The Biodiversity Riches of Africa"—aims to provide an overview of this incredibly diverse and biological important part of the world, exploring not only the flora and fauna found there and their conservation but also why this diversity is so important on a global scale. Throughout these sub-chapters, I hope you enjoy learning about the wonders of African biodiversity as much as I have and marvel at the wonderful and unique organisms and ecosystems found on this continent.

1.1.1 Overview of African Biodiversity

Biodiversity is important to the health of the Earth and the survival of humans. Human life depends on the Earth's two and a half million species of animals, plants, and microbes, and 20,000 different communities of people. All around the world, people need nourishing food and clean water for basic survival, but both are becoming scarce for some people. Recent medical research has shown that health products from natural sources are important in fighting diseases in countries globally. This means that continuing research into herbs and other medicinal plants occurs in a wide range of ecosystems is essential for scientific or medical progress. Even certain agricultural and forestry practices are dependent on biodiversity and have been used for centuries by people in order to survive.

Biodiversity is the variety of life on Earth. This includes the variety of species and ecosystems on the planet, whether terrestrial, marine, or other aquatic ecosystems or the diversity found within and between species and ecological communities

E. D. Wiafe, *Wildlife Conservation and Agriculture Development in Africa*, Sustainable Development Goals Series, https://doi.org/10.1007/978-3-032-09233-5_1

(Gaia 2024; WHO 2025). The term encompasses the variety of genetic information contained in the individual organisms within a species or populations of the species, including a variety of such characteristics as survival strategies and the ability to cope with environmental change, for example (Hoban 2021). In short, biodiversity is not just about the wide array of life forms, but the variety of Earth's environments in which they exist, and the interactions that support life. Biodiversity is a broad term and embraces a number of concepts. These include diversity within species, between species diversity, and of ecosystems (Gaia 2024; Biological Diversity Program nd). Thus, biodiversity is not just the number of different species of animals and plants—which often first come to mind—but encompasses variety at many levels. To a great extent, these concepts build on the idea of evolution and some basic ecological processes.

As the title indicates, in this part of the book the author sets the scene by providing an overview of the rich tapestry of life found in the continent of Africa. Firstly, Africa is shown to host an incredibly wide range of habitats. For instance, the book explains that these might be vast areas such as the Sahara Desert or smaller but just as distinctive habitats like the Giant Lobelia stands creating Actif volcano in Rwanda. In addition, the varied climates that are present in Africa contributes to the existence of varied life entities. For example, the intensely hot and dry areas in the north of the continent are reflected by typical desert species like the Fennec Fox. In contrast, the lush vegetation, large rivers and large diversity of life in rainforests are explained by Congo. The "big players" of the animal kingdom that you might first think of when considering African wildlife—namely elephants, giraffes and the big cats are as a result of combination of climatic, vegetational and other physical characteristics of Africa. There exist lesser known but just as impressive species such as the aardvark, the secretary bird and the naked mole rat coupled with the incredible variety of plant species in Africa. For example, baobab trees and their distinctive and ancient appearance are distinguished and so are the impressive varieties of bird species found on the continent.

1.1.2 Importance of African Biodiversity

It is vital to raise awareness of the numerous benefits of biodiversity and its vast importance in our lives. Although oftentimes overlooked, innumerable research and development projects exist all over the globe, with the aim of accessing the potential of biological diversity. Many have been focused on global food security and healthcare provision, but numerous other less common advantages have also been explored. For example, in the search for new pharmaceuticals, biological resources have been examined for their potential. All kinds of life are interlinked; a particular species may not seem particularly useful or important, but this may be linked to another species that is a key resource for biotechnology. In short, the study of the interconnections within such a rich variety of living organisms could provide the most significant scientific discoveries of the coming decades. This vast potential is what drives scientists and researchers to continue to study and utilize biological diversity. The importance of African biodiversity covers the vast and varied benefits to the people of the African continent. It spans from the potential for increasing food security, economic development, and environmental stability to climate change mitigation (Reuters 2025).

1.1.3 Threats to African Biodiversity

Despite the fact that biodiversity is so important to human life, we are destroying it at an unparalleled pace. Unfortunately, as the biodiversity of our planet continues to decrease, so does the possibility for sustainable development of its resources. A biological resources base that has been depleted by extinction provides fewer products and services, and the loss of species and ecosystems can never be recovered. Moreover, the indecision over how to best use and conserve Earth's resources will only become more complex as the human population continues to increase. This means that the need for biodiversity conservation is both immediate and long term (Chapman et al. 2022).

One major threat to biodiversity in Africa is the overexploitation of certain species. Many African species, including both plants and animals, are in high demand for international trade. This trade is driven by a variety of factors, including the demand for exotic pets, traditional Chinese medicine, and ivory and other animal products. As a result, species are often harvested, traded, and sold at unsustainable rates. For example, the African grey parrot, one of the most intelligent bird species, is highly sought after as a pet due to its longevity in captivity and its ability to mimic human speech. This has led to dramatic population declines in countries such as Ghana, where the species has been extensively trapped for the pet trade (African Center 2022). Similarly, large-scale poaching driven by international demand for ivory and rhino horn has had catastrophic impacts on elephants and rhinos in Africa. It's estimated that an elephant is killed every 15 min for its ivory. In some areas, this is leading to local extinctions and having serious ecological and physiological impacts on the populations as a whole. In the case of both the elephants and the rhinos, countries in Asia are the primary consumers of the illegal ivory and rhino horn. In recent years, conservationists have made huge strides in public policy change, and many countries, including the United States and China, have implemented near-total bans on the ivory trade. However, significant challenges remain and continued poaching for the illegal wildlife trade continues to devastate African populations.

This said, there are numerous cases when we still do not fully understand the role of a species or its importance in maintaining the structure and function of an ecosystem. Global environmental changes increase the risk of species becoming endangered or extinct as they are forced to move into areas with other species already using the resource, or into environments beyond their tolerance range. A loss of species would mean a decrease in biodiversity, and it is the loss of species that may have unwanted effects, or there may be "cryptic extinctions" in that the reduced populations would not be viable in the long term (Bellard et al. 2012).

This may sound essentially a trivial matter, but a loss of ecosystems could have a disastrous impact on the well-being of humans who are ultimately dependent on the resources provided by the biosphere for sustenance and economic wealth (Greenpeace 2024). Ecosystem sustainability is the ability to maintain the structure and functions of the ecosystem over long periods of time, and this is dependent on a species' ability to adapt to changes in the environment.

Extending this to a community level, if all species within the community are totally dependent on the same resources, the result would be unstable population sizes with many species experiencing varying periods of abundance or scarcity. There is a strong probability that they will experience abiotic events such as fire, disease, and biotic event like invasion by predators or other species, with the result being a shift from one type of community to another (i.e., transitions between alternative stable states).

To consider the importance of biodiversity, link the ideas of Gause for competition and competitive exclusion with the concept of an ecological niche. A follows B in that the more similar two species are in their requirements, the more overlap will occur in their niches and there will be interference in the ability of one to use the resource. In competition between the two, a decrease in numbers of one or both will result, or one will be out-competed and become extinct.

1.1.4 Africa's Unique Natural Heritage

Africa boasts an incredibly diverse array of natural wonders, from vast savannas to lush rainforests, towering mountains to stunning coastlines (Fig. 1.1). Its unique natural heritage encompasses a variety of ecosystems and species found nowhere else on Earth. Here are some highlights:

Serengeti Plains and Maasai Mara: The Serengeti ecosystem, stretching across Tanzania and Kenya, is renowned for its annual wildebeest migration, one of the most spectacular wildlife events on the planet. The Maasai Mara National Reserve in Kenya is part of this ecosystem and is home to an abundance of wildlife, including lions, elephants, and giraffes.

The Serengeti Plains and the Maasai Mara are two iconic landscapes in East Africa known for their breathtaking natural beauty and incredible wildlife. Here's a brief overview of each:

Located primarily in Tanzania, the Serengeti is one of the most famous savanna ecosystems in the world. It spans approximately 30,000 km^2 and is

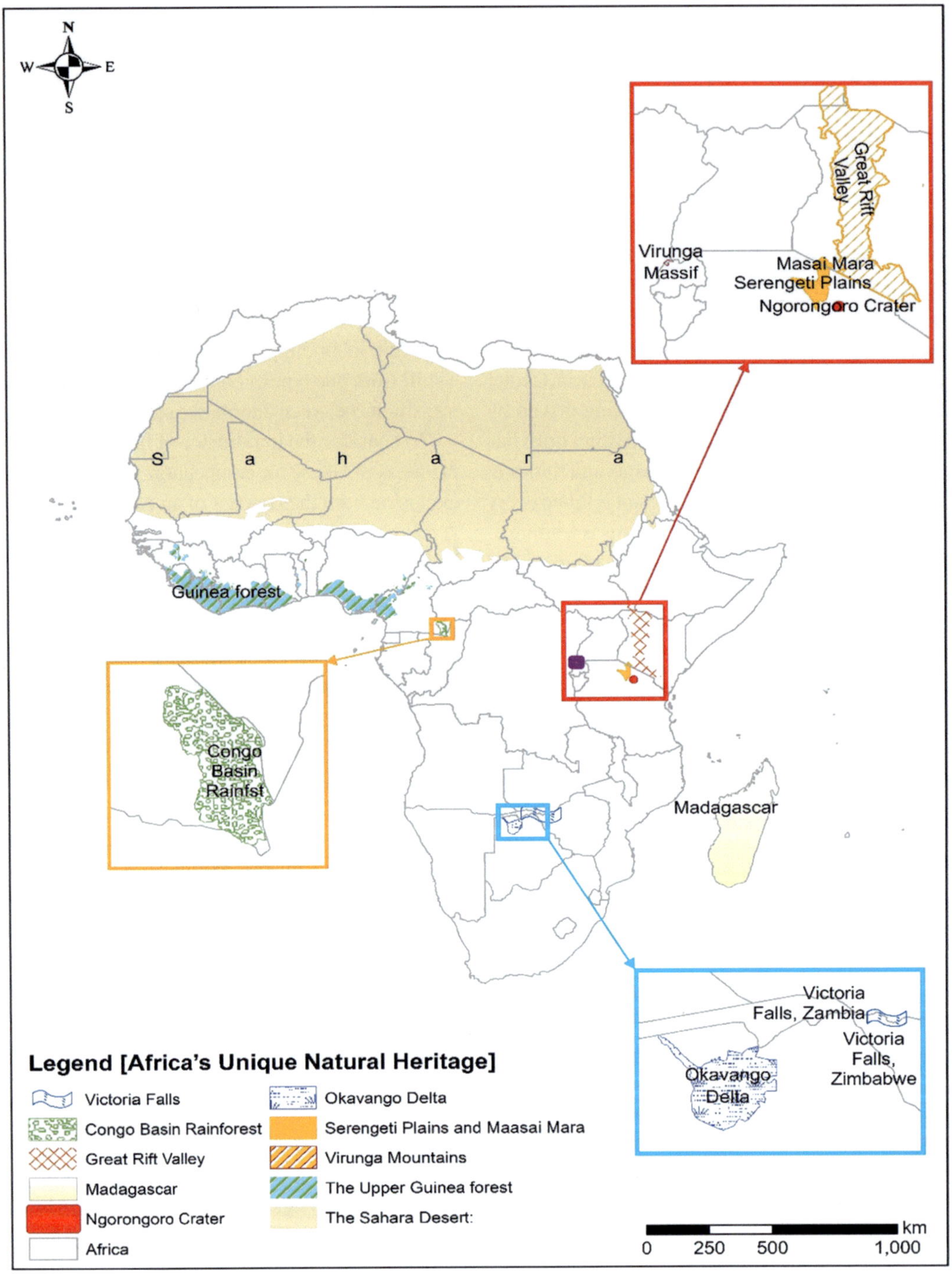

Fig. 1.1 Africa's unique natural heritage

characterized by vast grasslands, acacia woodlands, and scattered kopjes (rocky outcrops).

The Serengeti is renowned for its annual wildebeest migration, considered one of the most spectacular wildlife events on the planet. Each year, over two million wildebeest, along with hundreds of thousands of zebras and gazelles, migrate in search of fresh grazing lands, following a circular route that takes them through the Serengeti and into Kenya's Maasai Mara. The park is home to a diverse array of wildlife, including the Big Five (lion, elephant, buffalo, leopard, and rhinoceros), cheetahs, giraffes, hyenas, and numerous bird species.

The Serengeti also contains several conservation areas and reserves, such as the Ngorongoro Conservation Area and the Grumeti Game Reserve.

On the other hand, situated in southwestern Kenya, the Maasai Mara is an extension of the Serengeti ecosystem and shares its borders with Tanzania. It covers an area of about 1500 km^2.

The Maasai Mara is named after the Maasai people who inhabit the region and the Mara River, which flows through the reserve. The Maasai Mara is part of the larger Mara-Serengeti ecosystem, and it is known for its high density of wildlife, particularly during the migration season.

Like the Serengeti, the Maasai Mara hosts the annual wildebeest migration, with millions of animals crossing the Mara River from July to October, facing numerous challenges including crocodile-infested waters and predatory ambushes. Apart from the migration, the Maasai Mara offers excellent year-round game viewing opportunities, with abundant populations of predators and other wildlife species. The Maasai Mara is also home to the Maasai people, who continue to maintain their traditional pastoralist lifestyle and cultural practices amidst the wildlife-rich landscape.

Both the Serengeti Plains and the Maasai Mara are UNESCO World Heritage Sites and are critical for the conservation of biodiversity in East Africa. They attract tourists from around the world who come to witness the awe-inspiring spectacle of the African wilderness and to experience the rich cultural heritage of the region.

Okavango Delta: Located in Botswana, the Okavango Delta is the largest inland delta in the world. It supports a rich variety of wildlife and is a UNESCO World Heritage Site. The seasonal flooding of the delta creates a unique and constantly changing habitat, attracting numerous species of birds, mammals, and reptiles. The Okavango Delta is a stunning natural wonder located in Botswana, Southern Africa.

The Delta is formed by the Okavango River, which originates in the highlands of Angola. Instead of flowing into the sea, the river empties into the Kalahari Desert in Botswana, creating a unique inland delta. The delta covers an area of approximately 15,000 km^2, making it one of the largest inland deltas in the world.

The Okavango Delta is a mosaic of channels, lagoons, floodplains, and islands, which undergo seasonal flooding and drying, render it a unique ecosystem. This dynamic ecosystem supports an incredibly diverse array of wildlife, including numerous species of mammals, birds, reptiles, and fish. The delta is home to an abundance of wildlife, including iconic African species such as elephants, lions, leopards, buffalo, hippos, and crocodiles. It's also a haven for birdwatchers, with over 500 bird species recorded in the area, including various species of herons, storks, eagles, and waterfowl. The Okavango Delta is a premier safari destination, offering some of the best game viewing experiences in Africa. Visitors can explore the delta's pristine wilderness on guided game drives, walking safaris, and mokoro (traditional dugout canoe) excursions. There are also options for boat safaris and scenic flights over the delta, providing unique perspectives of this breathtaking landscape.

The Okavango Delta is a protected area, with significant efforts dedicated to conservation and sustainable tourism practices. It forms part of the larger Okavango Delta UNESCO World Heritage Site, recognized for its ecological significance and biodiversity. In addition to its natural beauty and wildlife, the Okavango Delta is also home to indigenous communities, including the Bayei and Hambukushu people, who have traditionally lived in harmony with the delta's environment. Visitors have the opportunity to learn about their cultures and traditions through cultural experiences and interactions.

Overall, the Okavango Delta is a remarkable destination that offers a perfect blend of wilderness adventure, stunning landscapes, and cultural richness, making it a bucket-list destination for nature enthusiasts and safari-goers alike.

Virunga Mountains: Straddling the borders of Rwanda, Uganda, and the Democratic Republic of Congo, the Virunga Mountains are home to the endangered mountain gorillas. These majestic creatures, made famous by Dian Fossey's work, are found only in this region and are a major draw for ecotourism.

The Virunga Mountains are a chain of volcanic mountains located in East Africa, spanning the borders of the Democratic Republic of Congo (DRC), Rwanda, and Uganda. The range is home to some of the last remaining mountain gorillas in the world and is renowned for its biodiversity and stunning landscapes.

The Virunga Mountains extend for about 80 km (50 miles) along the border region of the three countries. They are part of the Albertine Rift, a western branch of the East African Rift.

The mountains are volcanic in origin, with some of the peaks being dormant or extinct volcanoes. The highest peak is Mount Karisimbi, which reaches an elevation of 4507 m (14,787 feet) above sea level.

The Virunga Mountains are home to a rich diversity of flora and fauna. They are perhaps most famous for their population of critically endangered mountain gorillas, which are protected within several national parks in the region, including Virunga National Park in the DRC, Volcanoes National Park in Rwanda, and Mgahinga Gorilla National Park in Uganda.

Conservation efforts in the Virunga region are crucial for protecting the habitat of the mountain gorillas and other wildlife. However, the area has faced numerous challenges including armed conflicts, poaching, and habitat destruction.

Despite the challenges, tourism plays a significant role in the region's economy. Gorilla trekking is a major attraction, drawing tourists from around the world to experience the unique opportunity to observe these majestic creatures in their natural habitat.

The Virunga Mountains are also culturally significant, with several indigenous communities residing in the area. These communities have traditionally lived in harmony with the natural environment and play a role in conservation efforts and tourism activities.

Overall, the Virunga Mountains are a globally important and biologically diverse region that faces significant conservation challenges but also offers incredible opportunities for tourism and research.

Ngorongoro Crater: Situated in Tanzania, the Ngorongoro Crater is the world's largest intact volcanic caldera and a UNESCO World Heritage Site. The crater floor is home to a diverse array of wildlife, including lions, elephants, and flamingos, all living in close proximity.

The Ngorongoro Crater is a large volcanic caldera located in northern Tanzania, within the Ngorongoro Conservation Area. It is a UNESCO World Heritage Site and is often referred to as the "Eighth Wonder of the World" due to its stunning natural beauty and exceptional biodiversity.

The Ngorongoro Crater was formed around 2–3 million years ago when a large volcano, possibly as high as Mount Kilimanjaro, erupted and collapsed upon itself, forming a caldera. The crater floor covers an area of approximately 260 km^2 (100 mi^2), making it one of the largest intact calderas in the world.

The crater's rim rises to an elevation of around 2200 m (7200 feet) above sea level, while the crater floor descends to approximately 600 m (2000 feet) below the rim. The steep walls of the crater create a natural enclosure, resulting in a unique ecosystem with diverse habitats.

The Ngorongoro Crater is renowned for its exceptional wildlife density. It is home to an estimated 25,000 large animals, including iconic species such as lions, elephants, buffalo, rhinos, and wildebeests. The crater also supports a high concentration of predators, making it one of the best places in Africa for wildlife viewing.

The Ngorongoro Conservation Area, within which the crater is located, was established in 1959 to protect the region's wildlife and cultural heritage. The Maasai people, who have lived in the area for centuries, coexist with the wildlife and continue to practice traditional pastoralism within the conservation area.

Tourism is a significant economic activity in the Ngorongoro Conservation Area, with the crater being a major tourist attraction. Visitors can explore the crater floor on guided safari tours, observing the abundant wildlife and enjoying the breathtaking scenery. The conservation area also offers opportunities for cultural experiences, including visits to Maasai villages.

Like many conservation areas in Africa, the Ngorongoro Crater faces various challenges, including habitat degradation, human–wildlife conflict, and pressure from tourism activities. Sustainable management practices and community involvement are crucial for addressing these challenges and ensuring the long-term protection of this unique ecosystem.

Overall, the Ngorongoro Crater is a remarkable natural wonder and an important conservation area that offers visitors a once-in-a-lifetime opportunity to experience Africa's wildlife and landscapes up close.

Victoria Falls: Straddling the border between Zambia and Zimbabwe, Victoria Falls is one of the largest and most famous waterfalls in the world. Known as "The Smoke that Thunders," the falls are a UNESCO World Heritage Site and a breathtaking natural wonder.

Victoria Falls, one of the most famous natural wonders in the world, is a waterfall located on the Zambezi River at the border between Zambia and Zimbabwe in southern Africa. It is considered the largest waterfall in the world based on its combined width and height, making it a UNESCO World Heritage Site and a popular tourist destination.

Victoria Falls was formed as a result of the Zambezi River cutting through a series of basalt gorges over millions of years. The falls stretch across the width of the Zambezi River, forming a roughly 1700-m wide (5600-foot-wide) curtain of water that plunges into a deep gorge below.

The waterfall's height varies depending on the time of year and the volume of water flowing over it. At its highest point, Victoria Falls drops to around 108 m (354 feet), making it one of the largest waterfalls in the world in terms of height. During the rainy season, the falls are at their most powerful, with a massive volume of water cascading over the edge.

Name: The falls were named by the Scottish explorer David Livingstone in honor of Queen Victoria of the United Kingdom, who was reigning at the time of his discovery in 1855. However, the indigenous name for the falls is "Mosi-oa-Tunya," which means "the smoke that thunders" in the local Tonga language, a reference to the mist and roar created by the cascading water.

Biodiversity: The area surrounding Victoria Falls is rich in biodiversity, with lush rainforests and diverse wildlife habitats. The falls themselves support a unique ecosystem of plants and animals adapted to the misty conditions and turbulent waters. Visitors to the area can see a variety of bird species, monkeys, and other wildlife.

Tourism: Victoria Falls is a major tourist attraction, drawing visitors from around the world who come to marvel at its beauty and power. There are several viewpoints on both the Zambian and Zimbabwean sides of the falls, offering stunning panoramic views. Activities for tourists include guided tours, boat cruises, helicopter flights over the falls, and adventurous pursuits such as bungee jumping and white-water rafting.

Efforts are underway to protect and preserve the natural and cultural heritage of Victoria Falls and its surrounding ecosystems. Sustainable tourism practices, environmental education programs, and conservation initiatives aim to minimize the impact of human activities on the fragile environment while maximizing the benefits to local communities.

Overall, Victoria Falls is not only a breathtaking natural wonder but also a symbol of the beauty and power of the natural world, attracting visitors with its awe-inspiring scenery and diverse array of experiences.

The Sahara Desert: Covering much of North Africa, the Sahara is the largest hot desert in the world. Its vast expanse of sand dunes, rocky plateaus, and oases is home to a unique array of flora and fauna adapted to its extreme conditions.

The Sahara Desert is the largest hot desert in the world, covering an area of approximately 9.2 million km^2 (3.6 million mi^2). It spans across North Africa, stretching from the Atlantic Ocean in the west to the Red Sea in the east, and from the Mediterranean Sea in the north to the Sahel region in the south.

The Sahara Desert encompasses a diverse range of landscapes, including sand dunes, rocky plateaus, gravel plains, and mountain ranges such as the Atlas Mountains in the north and the Tibesti Mountains in the central Sahara. It is characterized by its arid climate, extreme temperatures, and limited vegetation cover.

The Sahara experiences a hyper-arid climate, with very low annual rainfall and high temperatures. Daytime temperatures can soar to over 50°C (122°F) in the summer months, while nighttime temperatures can drop below freezing in the winter. Rainfall is sporadic and generally occurs in brief, intense bursts.

Despite its harsh conditions, the Sahara supports a surprising variety of plant and animal life adapted to desert environments. Vegetation is sparse and mainly consists of drought-resistant plants such as acacia trees, cacti, and succulents. Animal species found in the Sahara include desert-adapted mammals like camels, fennec foxes, and addax antelopes, as well as numerous reptiles, birds, and insects.

The Sahara has a rich history of human occupation dating back thousands of years. It was once home to ancient civilizations such as the Egyptians, Berbers, and Tuareg people, who developed sophisticated societies and trading networks despite the challenges of the desert environment. The Sahara also played a significant role in the trans-Saharan trade routes, connecting sub-Saharan Africa with the Mediterranean world.

The Sahara Desert faces various environmental and socio-economic challenges, including desertification, drought, and resource scarcity. Desertification, driven by factors such as deforestation, overgrazing, and climate change, threatens to expand the desert's boundaries and degrade fragile ecosystems. Additionally, communities living in the Sahara must contend with water scarcity, food insecurity, and economic marginalization.

Despite its inhospitable nature, the Sahara attracts adventurous travelers and researchers interested in exploring its unique landscapes, cultural heritage, and natural wonders. Popular activities for tourists include camel trekking, desert safaris, and visits to ancient archaeological sites and traditional Berber villages. Overall, the Sahara Desert is a vast and fascinating wilderness that continues to captivate the imagination of people around the world, offering both challenges and opportunities for those who venture into its harsh but beautiful terrain.

Madagascar: This island nation off the eastern coast of Africa is a biodiversity hotspot, with a staggering number of species found nowhere else on Earth. From the iconic lemurs to bizarre chameleons and unusual baobab trees, Madagascar's flora and fauna are truly unique.

Madagascar is an island country located in the Indian Ocean, off the southeastern coast of Africa. It is the fourth largest island in the world and is known for its unique biodiversity, stunning landscapes, and vibrant culture.

Madagascar is characterized by diverse landscapes, including lush rainforests, arid deserts, pristine beaches, and towering mountains. The island is home to a wide range of ecosystems, from tropical rainforests in the east to spiny forests in the south. Madagascar is often referred to as a "biodiversity hotspot" due to its remarkable array of unique plant and animal species found nowhere else on Earth. The island is famous for its lemurs, a type of primate endemic to Madagascar, as well as numerous species of chameleons, baobab trees, and other flora and fauna.

The culture of Madagascar is a blend of African, Asian, and European influences, reflecting the island's diverse history of settlement and trade. The Malagasy people have a rich cultural heritage, with unique traditions, music, dance, and cuisine. Traditional practices such as ancestor worship and famadihana (the turning of the bones ceremony) are still observed in many parts of the island. Madagascar was settled by seafaring peoples from Southeast Asia around 2000 years ago. Over the centuries, the island was also visited by Arab and Persian traders, as well as European explorers. In the nineteenth century, Madagascar was colonized by the French, and it gained independence in 1960. Agriculture is the backbone of Madagascar's economy, with the majority of the population engaged in subsistence farming. The island is known for its production of vanilla, cloves, and other spices, as well as exports of coffee, cocoa, and rice (Fig. 3.5). Tourism is also an important economic sector, with visitors drawn to Madagascar's natural beauty and unique wildlife.

Madagascar faces significant environmental challenges, including deforestation, habitat loss, and biodiversity loss. Efforts are underway to conserve the island's unique ecosystems and protect endangered species, with the establishment of national parks and reserves and initiatives to promote sustainable development and ecotourism.

Overall, Madagascar is a fascinating and enchanting destination, offering visitors a glimpse into a world of unparalleled natural beauty, rich cultural heritage, and extraordinary biodiversity.

Great Rift Valley: Stretching from the Middle East to Mozambique, the Great Rift Valley is a geological marvel and a hotspot of biodiversity. It encompasses diverse landscapes, including lakes, mountains, and grasslands, and is home to a wide variety of plant and animal species. The Great Rift Valley is a vast geographic trench system that stretches approximately 6000 km (3700 miles) from the Beqaa Valley in Lebanon to Mozambique in East Africa. It is one of the most distinctive geographical features on Earth and is known for its stunning landscapes, diverse ecosystems, and rich geological history.

The Great Rift Valley is composed of a series of interconnected rifts, faults, and depressions that run along the tectonic plate boundaries of the African and Arabian plates. It encompasses a wide range of habitats, including savannas, forests, lakes, and volcanoes. The formation of the Great Rift Valley began around 35 million years ago as tectonic forces caused the Earth's crust to stretch and thin, leading to the gradual separation of the African and Arabian plates. Over time, this process created the deep valleys, escarpments, and volcanic peaks that characterize the Rift Valley landscape.

The Rift Valley is dotted with a series of lakes, many of which are freshwater and saline. These lakes vary in size and depth and provide important habitats for a diverse array of plant and animal species. Some of the most famous lakes in the Rift Valley include Lake Tanganyika, Lake Malawi, Lake Victoria, and the saline Lake Turkana.

The Great Rift Valley is renowned for its high levels of biodiversity, with numerous species of mammals, birds, reptiles, and plants inhabiting its diverse habitats. It is home to iconic African wildlife such as elephants, lions, giraffes, and rhinoceroses, as well as a variety of endemic species found nowhere else on Earth.

The Great Rift Valley is also rich in cultural heritage, with evidence of human occupation dating back millions of years. Archaeological sites within the Rift Valley have yielded fossils of early hominids, including the famous Australopithecus fossils discovered in the Olduvai Gorge in Tanzania. The region is also home to diverse ethnic groups and cultures, each with its own traditions, languages, and customs. The Great Rift Valley is a popular tourist destination, attracting visitors with its breathtaking scenery, wildlife viewing opportunities, and cultural attractions. Activities for tourists include safari tours, birdwatching, hiking, hot air balloon rides, and visits to archaeological sites and traditional villages.

Overall, the Great Rift Valley is a remarkable natural wonder and a testament to the dynamic forces that shape the Earth's surface. Its geological, ecological, and cultural significance make it a truly awe-inspiring destination for travelers and a vital area for scientific research and conservation efforts.

Congo Basin Rainforest: The second-largest tropical rainforest in the world, after the Amazon, the Congo Basin is a vital ecosystem that supports an incredible diversity of plant and animal life, including forest elephants, chimpanzees, and countless species of birds and insects.

The Congo Basin Rainforest, also known as the Congo Rainforest or Congo Basin Forest, is the second-largest tropical rainforest in the world, after the Amazon Rainforest. It spans across six countries in Central Africa, including the Democratic Republic of the Congo (DRC), Republic of the Congo, Cameroon, Central African Republic, Equatorial Guinea, and Gabon. Here are some key features and aspects of the Congo Basin Rainforest: The Congo Basin Rainforest covers an area of approximately 2 million km^2 (800,000 square miles), making it the world's second-largest tropical rainforest. It extends from the Atlantic Ocean in the west to the Albertine Rift in the east, and from the northern savannas to the southern rainforests of the Congo River basin.

The Congo Basin Rainforest is one of the most biodiverse regions on Earth, with an estimated 10,000 species of plants, 400 species of mammals, 1000 species of birds, and 700 species of fish. It is home to iconic African wildlife such as forest elephants, gorillas, chimpanzees, okapis, and various species of antelopes, monkeys, and birds.

The rainforest provides a wide range of ecosystem services that are essential for both local communities and the global environment. These services include carbon sequestration, regulation of the water cycle, provision of habitat for wildlife, and sustenance for indigenous peoples who rely on the forest for food, medicine, and cultural practices. The Congo Basin Rainforest faces numerous threats, including deforestation, habitat fragmentation, illegal logging, mining, infrastructure development, and poaching. These activities are driven by factors such as population growth, agricultural expansion, and demand for timber, minerals, and other natural resources.

Efforts to conserve the Congo Basin Rainforest are underway, led by governments, non-governmental organizations (NGOs), and international initiatives. Protected areas such as national parks, reserves, and community-managed forests are established to safeguard biodiversity and promote sustainable land management practices. Initiatives to combat illegal logging, promote sustainable forestry, and support indigenous rights are also critical for conservation efforts in the region.

The Congo Basin Rainforest plays a crucial role in mitigating climate change by absorbing and storing vast amounts of carbon dioxide from the atmosphere. However, deforestation and degradation of the forest can release carbon into the atmosphere, contributing to global warming. Therefore, protecting and restoring the rainforest is essential for climate change mitigation efforts. Overall, the Congo Basin Rainforest is a vital natural resource with immense ecological, cultural, and economic significance. Preserving this unique ecosystem is essential for biodiversity conservation, climate change mitigation, and the well-being of both local communities and the global environment.

The Upper Guinea forest: The Upper Guinea forest is a significant forest ecosystem located in West Africa, spanning across several countries along the Guinea Coast, including Guinea, Sierra Leone, Liberia, Ivory Coast (Côte d'Ivoire), Ghana, and Togo. It is one of the 25 biodiversity hotspots in the world, characterized by its rich flora and fauna, unique ecological features, and cultural importance.

The Upper Guinea forest is renowned for its exceptional biodiversity, with high levels of species endemism and diversity. It is home to a wide range of plant and animal species, including many that are found nowhere else on Earth. The forest supports a variety of habitats, including tropical rainforests, gallery forests, mangroves, and savannas.

The forest is characterized by dense, lush vegetation dominated by a variety of tree species, including mahogany, iroko, ebony, and rubber trees. It is also home to numerous plant species with medicinal properties, which are used by local communities for traditional healing practices.

The Upper Guinea forest is home to a diverse array of wildlife, including mammals such as chimpanzees, forest elephants, pygmy hippos, and various species of monkeys and antelopes. It is also a haven for birdwatchers, with over 700 species of birds recorded in the region, including many endemic and migratory species.

The Upper Guinea forest faces significant threats from human activities, including deforestation, agricultural expansion, logging, mining, and infrastructure development. These activities are driven by factors such as population growth, poverty, and demand for timber, minerals, and agricultural land. Habitat fragmentation and degradation also pose serious threats to the forest's biodiversity and ecological integrity.

Efforts to conserve the Upper Guinea forest are underway, led by governments, conservation organizations, and local communities. Protected areas such as national parks, reserves, and community-managed forests are established to safeguard biodiversity and promote sustainable land management practices. Initiatives to combat illegal logging, promote sustainable forestry, and support community-based conservation are also critical for conservation efforts in the region.

The Upper Guinea forest has cultural significance for the indigenous communities living in the region, who rely on the forest for food, medicine, shelter, and cultural practices. Traditional knowledge and practices related to forest management and conservation are passed down through generations and play an important role in sustainable resource use. Overall, the Upper Guinea forest is a vital ecosystem with immense ecological, cultural, and economic value. Preserving this unique forest is essential for biodiversity conservation, climate change mitigation, and the well-being of both local communities and the global environment.

These are just a few examples of Africa's unique natural heritage, which continues to awe and inspire visitors from around the world with its beauty, diversity, and ecological significance.

1.1.5 The Role of Wildlife in Ecosystems

The role of wildlife in ecosystems in Africa is crucial and multifaceted, contributing to the overall health, balance, and functioning of these complex systems. Here are some key aspects of the role of wildlife in African ecosystems: Wildlife species in

Africa represent a diverse array of plants, animals, and microorganisms that inhabit various habitats, including savannas, forests, wetlands, deserts, and coastal areas. This biodiversity is essential for maintaining ecosystem resilience and adaptability to environmental changes (National Geographic Education 2025).

Wildlife play a critical role in providing ecosystem services that benefit both humans and the environment. These services include pollination, seed dispersal, nutrient cycling, soil formation, water purification, and pest control (WHO 2025). For example, pollinators such as bees and butterflies are essential for the reproduction of many plant species, including food crops.

Wildlife species interact with each other and with their environment through complex food webs and ecological relationships. Predators help regulate prey populations, preventing overgrazing and maintaining ecosystem balance. Herbivores influence vegetation dynamics through grazing and browsing, shaping plant communities and habitats.

Many wildlife species, such as birds, mammals, and insects, play a crucial role in seed dispersal by consuming fruits and seeds and dispersing them over long distances. This process promotes plant regeneration, genetic diversity, and the colonization of new habitats, contributing to ecosystem resilience and adaptation to environmental change (Fig. 1.2).

Some wildlife species act as habitat engineers, modifying their environment in ways that create or enhance habitats for other species. For example, elephants are known to create pathways through dense vegetation, dig water holes during dry seasons, and topple trees, creating openings in the forest canopy that promote plant regeneration and diversity (National Geographic Education 2025).

Wildlife in Africa holds significant cultural and socioeconomic value for local communities, who often depend on wildlife resources for food, medicine, clothing, and cultural practices. Ecotourism based on wildlife viewing and conservation is also a major source of revenue and employment in many African countries, contributing to local economies and livelihoods.

Despite their importance, African wildlife and ecosystems face numerous threats, including habitat loss and degradation, poaching, illegal wildlife trade, climate change, pollution, and human–wildlife conflict. Conservation efforts are crucial for protecting and restoring wildlife populations and their habitats, ensuring the long-term sustainability of African ecosystems and the benefits they provide to both humans and the environment.

In summary, wildlife in Africa plays a fundamental role in maintaining the health, resilience, and functioning of ecosystems, while also providing cultural, socioeconomic, and ecological benefits to local communities and the broader environment. Protecting and conserving African wildlife and their habitats is essential for ensuring the continued well-being of both humans and biodiversity in the region.

1.1.6 Approaches to Biodiversity Conservation

The knowledge that we gain through these and other methods must be applied if policies and attitudes are to be changed both at the level of the general public and at the level of individual decision-makers. Changes in policy and public opinion are crucial if we are to successfully conserve biodiversity.

There are a number of techniques and methods that may turn out to be useful in conserving biodiversity in the future. One vital method is the development of predictive models that will allow us to assess the impact of different future scenarios on the environment and its various components. This tool will be of use to policymakers in making informed decisions when creating laws and regulations that may impact the environment. Computer technology has also revolutionized the way we can gather and process information. Development in this area has led to improvements in the ability of scientists to monitor and access changes in the environment over long- and short-time scales. This is crucial if we are to act as stewards in the wise management of natural resources.

Species, coupled with variety and variability of plants, animals, and microorganisms, represent a vast resource from which humanity has drawn biological, cultural, and aesthetic wealth. The extinction of species and loss of biological diversity

Fig. 1.2 Pictures of savannah elephants (*Loxodonta africana*) (**A**–**C**); young bushbuck (*Tragelephus scriptus*) (**D**); patas monkey (*Erythrocebus patas*) (**E**); Warthog (*Phacochoerus africanus*) (**F**) on African savannah landscapes. [Photo by E. A. Tetteh]

is likely to proceed further rapidly in the future. The intricate, dynamic, and complex interactions between the components of ecological systems mean that the damage is often irreversible. This demands action to take place at every level from local to global. Because of the complexity of the issue, the solutions will require the use of innovative approaches (Table 1.1).

Fig. 1.2 (Continued)

Fig. 1.2 (Continued)

Table 1.1 List of Protected Areas in Africa with iconic species, conservation efforts, threats, and significance

Protected area	Country	Iconic wildlife species	Conservation efforts	Threats	Tourism/ biodiversity significance
W National Park	Niger, Burkina Faso, Benin	West African lion, cheetah, elephant, African buffalo	UNESCO Biosphere Reserve, anti-poaching programs	Poaching, habitat loss due to agriculture	Key biodiversity hotspot for Sahelian ecosystems
Mole National Park	Ghana	African elephant, kob antelope, hyena	Community-based conservation, eco-tourism initiatives	Human–wildlife conflict, illegal hunting	Largest national park in Ghana, home to rare mammals
Tai National Park	Côte d'Ivoire	Pygmy hippo, chimpanzee, African golden cat	UNESCO World Heritage Site, primate conservation programs	Logging, agricultural encroachment	One of the last primary rainforests in West Africa
Comoé National Park	Côte d'Ivoire	Elephant, leopard, giant pangolin	UNESCO-listed, patrols against poaching	Illegal poaching, habitat destruction	High biodiversity with savanna-forest transition zone
Bui National Park	Ghana	Hippopotamus, African buffalo, roan antelope	Hippo sanctuary, protected river zones	Dam construction, habitat fragmentation	Significant aquatic biodiversity
Dzanga-Sangha Special Reserve	Central African Republic	Forest elephant, western lowland gorilla, bongo	WWF-supported conservation, eco-tourism	Deforestation, poaching for ivory and bushmeat	Important site for gorilla tracking
Odzala-Kokoua National Park	Republic of Congo	Western lowland gorilla, forest buffalo, chimpanzee	Anti-poaching patrols, sustainable tourism programs	Illegal logging, mining, poaching	One of Africa's best-preserved rainforests
Salonga National Park	Democratic Republic of Congo	Bonobo, Congo peafowl, African golden cat	UNESCO-listed, research programs for bonobo protection	Armed conflict, illegal hunting	Largest tropical rainforest park in Africa
Tassili n'Ajjer National Park	Algeria	Barbary sheep, fennec fox, dorcas gazelle	UNESCO site, rock art preservation	Climate change, illegal hunting	Rich in prehistoric rock art and desert biodiversity
Banc d'Arguin National Park	Mauritania	Dolphins, flamingos, green sea turtle	Ramsar Wetland, marine protection programs	Overfishing, pollution	Key habitat for migratory birds and marine life
Serengeti National Park	Tanzania	Wildebeest, African lion, cheetah, elephant	Anti-poaching patrols, Great Migration monitoring	Poaching, human–wildlife conflict	Hosts the Great Migration, a major wildlife spectacle
Maasai Mara National Reserve	Kenya	Cheetah, African elephant, wildebeest, lion	Community conservation efforts with Maasai people	Habitat destruction, over-tourism	World-famous for lion populations and migration events

(continued)

Table 1.1 (Continued)

Protected area	Country	Iconic wildlife species	Conservation efforts	Threats	Tourism/ biodiversity significance
Virunga National Park	DRC	Mountain gorilla, chimpanzee, African fish eagle	UNESCO site, strong anti-poaching operations	Armed conflicts, illegal charcoal trade	One of the last habitats for mountain gorillas
Ngorongoro Conservation Area	Tanzania	Black rhino, spotted hyena, flamingos	UNESCO-listed, Maasai conservation integration	Overgrazing, tourism pressure	Home to Ngorongoro Crater, a unique volcanic ecosystem
Kruger National Park	South Africa	African leopard, white rhino, African buffalo	Advanced anti-poaching units, rhino protection efforts	Rhino poaching, human encroachment	One of the largest game reserves in Africa
Etosha National Park	Namibia	Black rhino, springbok, African lion	Waterhole conservation projects, black rhino protection	Climate change, habitat fragmentation	Famous for salt pan landscapes and wildlife diversity
Chobe National Park	Botswana	African elephant, Nile crocodile, giraffe	Strong elephant conservation programs	Drought, illegal wildlife trade	Home to one of the largest elephant populations in Africa

Source: Compiled from UNESCO World Heritage Centre, Ramsar Convention Secretariat, WWF Reports, and IUCN Red List Data, 2015–2023

References

African Biodiversity Programme (nd) Biodiversity: diversity within species, between species, and of ecosystems. African Biodiversity Programme

Archer E, Dziba L, Jo Mulongoy K, Malebajoa Anicia Maoela MA, Walters M, Biggs R et al (2018) Biodiversity and ecosystem services for Africa. IPBES Regional Assessment Report. https://files.ipbes.net/ipbes-web-prod-public-files/spm_africa_2018_digital.pdf

African Center (2022) African biodiversity loss raises risk to human security. African Center for Strategic Studies. https://africacenter.org/spotlight/african-biodiversity-loss-risk-human-security/

Bedair H (2023) Global warming status in the African continent. Environ Sci Pollut Res 30:12345–12360. https://doi.org/10.1007/s41742-023-00534-w

Bellard C et al (2012) Impacts of climate change on the future of biodiversity. Ecol Lett 15(4):365–377. https://doi.org/10.1111/j.1461-0248.2012.01736.x

Chapman CA, Abernathy K, Chapman LJ, Downs C, Effiom EO et al (2022) The future of sub-Saharan Africa's biodiversity in the face of global challenges. Front Ecol Evol 10:790552. https://doi.org/10.3389/fevo.2022.790552

Gaia Company (2024) The different types of biodiversity (+ examples). https://gaiacompany.io/the-different-types-of-biodiversity/

Greenpeace (2024) What is biodiversity and why is it important? https://www.greenpeace.org/mena/en/what-is-biodiversity-and-why-is-it-important/

Hoban S (2021) No more trouble: an economic strategy to protect taxonomic, functional and phylogenetic diversity of continental turtles. Biol Conserv 261:109241. https://doi.org/10.1016/j.biocon.2021.109241

National Geographic Education (2025) Biodiversity. https://education.nationalgeographic.org/resource/biodiversity/

Reuters (2025) Can Africa use biodiversity credits to turn its rich carbon stores into hard cash? Reuters. https://www.reuters.com/sustainability/land-use-biodiversity/can-africa-use-biodiversity-credits-turn-its-rich-carbon-stores-into-hard-cash-2025-01-02/

UNESCO World Heritage Centre, Ramsar Convention Secretariat, World Wide Fund for Nature (WWF), & International Union for Conservation of Nature (IUCN) (2015–2023) Data on African protected areas, iconic wildlife species, conservation efforts, and threats. Retrieved from https://www.unesco.org,https://www.ramsar.org, https://wwf.panda.org,https://www.iucnredlist.org

WHO (2025) Biodiversity. World Health Organization. https://www.who.int/news-room/fact-sheets/detail/biodiversity

2 Human–Wildlife Conflict

2.1 Understanding Human–Wildlife Conflict

The impacts of human–wildlife conflict can be felt at both individual and community level and may have serious implications on the conservation of the species and the future of wildlife in areas where conflict is prevalent. Understanding the different forms of human–wildlife conflict, and the factors influencing the initiation and perpetuation of these conflicts, is essential in developing methods to mitigate conflict and in planning the best course of action to prevent future events. In order to understand why wildlife cause problems for people, and what can be done to minimize the occurrence of conflict in the future, conflict analysis is necessary which can provide systematic assessments of the conflicts that are occurring, and offer solutions to the conflicts that are sustainable in the long term for wildlife, and are socially and culturally appropriate for the people involved.

Conflict between people and animals exists in all parts of the world, and may only be a minor problem with a few individual animals, or a major issue involving whole populations. Human–wildlife conflict is any contact between humans and wildlife that causes negative impacts on human social, economic or cultural life, on the conservation of wildlife populations, or on the environment. This includes conflicts with wildlife that result in changes to the use of the land by humans, changes to the status of the species or, indeed, the removal of the "problem" animals by both legal and illegal means.

From these two understandings of conflict, human–wildlife conflict can be simplified as an interaction between humans and wildlife that leads to negative impacts, where humans see the wildlife as the troublemaker and require corrective measures to resolve the conflict. This definition is somewhat restrictive in terms of assigning a negative impact from wildlife to humans, as there have been cases where humans are the ones who start the conflict with wildlife, or vice versa. Lemer and Conover have simplified the definition into a broader aspect, stating that the conflict involves a competition for resources and differing values, which is more applicable to a wide variety of situations involving humans and wildlife. From this broader definition, conflict can result in a wildcard situation where neither party perceives harm to the other, but preventive measures are taken by one party against the other, and in extreme cases, it could cause death to one or the other.

Conflicts or competition between humans and wildlife have emerged as a result of the competition for shared resources between humans and wildlife, as well as differences in value systems attached to these resources (Conover 1998). Conflict is a social relationship that exists when two or more parties have the potential to exercise power in order to produce the least costly outcome for the other party.

E. D. Wiafe, *Wildlife Conservation and Agriculture Development in Africa*, Sustainable Development Goals Series, https://doi.org/10.1007/978-3-032-09233-5_2

Wildlife conflict experts define wildlife damage as any situation in which humans perceive wildlife as interfering with normal activities or needs, resulting in mental, physical, or both mental and physical harm to the person (Conover 2001).

The continued existence of all species (wildlife and human) depends on resources. Allocation of scarce, limiting resources required by one group often results in resource depletion and/or an alteration in the quality of these resources used by another group. The resources may be abiotic (e.g., water, habitat) or biotic (e.g., prey, livestock) and the "depletion or alteration" effect may be positive or negative. This competition or "contest interactions" often leads to conflict and for wildlife, this may result in wide-ranging effects—a complete breakdown of society with a resulting change in community structure and resource availability. When this type of interaction occurs between humans and wildlife and the result is a perceived negative effect on the part of the human, it is termed Human–Wildlife Conflict. This is a complex issue that has far-reaching impacts on a wide range of stakeholders and often results in a no-win scenario for one of the groups involved. This assignment will attempt to critically analyze the nature of contest interactions and the resulting conflict between humans and wildlife and assess the factors determining the severity of such conflicts and the potential for mitigation strategies to resolve the issues. This is an excerpt from an assignment that aims to analyze, in depth, the complexities of Human–Wildlife Conflict—an increasingly problematic issue in today's rapidly developing world. The focus and scope of the assignment has led to the decision to use case studies of African elephants in competition/coexistence with humans in Sri Lanka and African Buffalo with humans in the Lower Shire Valley of Malawi. This approach will allow for the comprehensive examination and synthesis of data about the elephants and humans, and the buffalo and humans. The first step in achieving our goal is to understand and describe the nature of these interactions and the resulting conflicts between the species involved. Understanding the type, form, and severity of the conflicts will allow us to discern the potential implications and effective mitigation strategies to alleviate the conflicts and promote coexistence.

2.2 The Impact of Wildlife on Agriculture

Agriculture forms the backbone of the African economy, which significantly contributes to that of the world. It supports the livelihoods of 80% of the population, supplies food to urban areas and raw materials to agro-industries, and generates export earnings. In the context of agricultural growth and poverty reduction, there is a need for a suitable environment and land for agriculture. Environmental constraints like land degradation, soil nutrient mining, deforestation, wildlife raiding and shortages of water and common property resources continue to threaten the sustainability of poor people's livelihoods. Land and tree planting investments have been attracting large public and private sector investments in Africa, publishing the fact that African agriculture is one of the fundamental drivers of global frontier expansion.

This has meant that most of the impact on the environment from agriculture, whether it be negative or positive, has typically been viewed with a single focus on the goal of producing more food and resources.

The result of this large expansion in agriculture is that it has been a major cause of deforestation, loss of natural grasslands, and various changes to many of the world's other ecosystems. Although agriculture has been a major cause of habitat change, it has also been a primary source of food and various other resources for the human population. With the human population still growing at an exponential rate, the need for more agricultural land and resources from the environment will continue to rise.

In the case of agriculture, the conversion of land from natural plant communities or other land uses into agricultural land has been widespread and has been one of the most dramatic changes to the Earth's system. Croplands cover around 12% of the Earth's ice-free land surface, while the total area sown to crops is about 1.5 times that amount. Land used for agriculture is now equal to the total area of North and South America combined.

Humans have been altering the environment for their own benefit since they first evolved. Human evolution can be seen as a continuing process of resource intensification, that is, humans have continually attempted to extract more resources from the environment around them. This has resulted in many changes to not just the local environment, but also global change.

2.2.1 Importance of Agriculture in Africa

Agriculture is the backbone of Africa's economy, contributing about 30% to the Gross Domestic Product (GDP) and employing 60%–70% of the population. The need to increase food production and the role of modernization as a means to this end have been recurrent themes in agricultural development in Africa. The primary objective of increasing food production is to feed the growing population and to alleviate rural poverty. This can be observed from the various projects, plans, and policies formulated by African nations and the various international development agencies over the years. For instance, the Comprehensive Africa Agriculture Development Programme (CAADP) has a goal of at least 6% annual growth in the agricultural sector. This is to be obtained through the allocation of at least 10% of the total national budget to agriculture and rural development over 10 years. The Programme for Africa's Seed Systems (PASS) aims to reform seed systems in Africa. The SADC Regional Indicative Strategic Development Plan seeks to increase agricultural production by producing a food surplus by 2020, and the list goes on. This commitment and determination to increase food production and ultimately improve the lives of Africans indicate the importance of agriculture in Africa, both as a source of livelihood and as a tool for development. The role of modernization in African agriculture has long been a topic of debate. The proponents of modernization emphasize the need to increase productivity in order to meet the food needs of the growing population, release labor to the industrial sector, provide raw materials for agro-industry, and to generate foreign exchange for importing essential goods. This is to be achieved through a shift from "traditional" to "modern" farming characterized by the use of modern inputs, machinery, and improved crop varieties. The many success stories of Asian and Latin American countries' transition out of poverty through agricultural development have provided the impetus to follow a similar path. An example of a modernization initiative in Africa was the Green Revolution for Africa (GRA) campaign headed by the Alliance for a Green Revolution in Africa (AGRA). This came after the widely acknowledged failure of a similar initiative in the form of the Consultative Group on International Agricultural Research (CGIAR) research centers in Africa. The GRA placed heavy emphasis on the use of improved seeds, fertilizers, and policy support as a means to African agricultural growth and poverty reduction. Although the GRA received mixed reviews, the notion of modernization as a means to agricultural development remains relevant in Africa.

The French broadened the meaning in early nineteenth century to include "a struggle between two species for a share of a common resource in limited supply". This is probably the most useful definition for the purposes of this book, as the common resource in question is habitat and food resources in agricultural areas, and limited supply infers that such resources are increasingly scarce due to loss of natural habitat. This definition also implies that conflict is a normal and regular aspect of human/wildlife interactions, and is an implicit recognition of the fact that wildlife is often a source of objective or subjective value.

A common fallacy concerning wildlife damage to agriculture is the belief that such problems are recent and increasing in nature. In actual fact, human/wildlife conflict is as old as agriculture itself. With domestication of plant and animals some 10,000 years ago, crop-raiding and competition with wild herbivores would have presented problems for early farming communities. Over the centuries, many different societies would have had to develop strategies to mitigate such conflict, and it could be that some traditional methods still employed today may be relevant to current problems.

2.3 Case Studies of Human–Wildlife Conflicts from Africa

Human–wildlife conflicts are a pervasive issue in Africa, causing significant challenges for both human communities and wildlife conservation efforts (Muhumuza and Balkwill 2013). The information collected in the study revealed that middle-aged respondents were the main contributors to the data. These respondents shared their views on the awareness level and perceptions towards wildlife conservation, indicating a concern for the effects of human–wildlife conflict impact on livelihoods (Fig. 2.1).

The impact of human–wildlife conflicts on local livelihoods is significant. Many communities in Africa rely on agriculture for their sustenance, and crop damage caused by wildlife can have devastating effects. Moreover, attacks on livestock by predators can lead to economic losses for herders and farmers. These conflicts not only threaten the food security of local communities but also contribute to increased poverty levels.

In Africa, there are multiple factors contributing to human–wildlife conflicts. The expansion of human settlements and agricultural activities into wildlife habitats has led to increased encounters between humans and wildlife. Additionally, climate change and environmental degradation have forced wildlife to seek food and water sources in areas populated by humans, further exacerbating the potential for conflicts.

Moreover, the illegal wildlife trade and poaching have disrupted natural ecosystems, resulting in imbalances in predator-prey dynamics and pushing some species closer to human-inhabited areas in search of food. Human activities such as logging, mining, and infrastructural development have also contributed to habitat loss for wildlife, leading to increased competition for resources and territorial conflicts.

The complexity of these factors highlights the need for comprehensive and sustainable solutions to mitigate human–wildlife conflicts while ensuring the conservation of Africa's diverse wildlife. Understanding the root causes of these conflicts is essential in developing effective strategies that balance the needs of local communities with the preservation of biodiversity. The conflicts between humans and wildlife in Africa have far-reaching consequences for both local communities and the conservation of biodiversity (Fig. 2.1).

Some examples of human–wildlife conflicts in Africa include:

- Elephants raiding crops in rural farming communities, causing significant damage and economic loss.
- Lions preying on livestock, leading to retaliatory killings by herders to protect their livelihoods.

Fig. 2.1 A group of Elephants in a raiding spree on farms adjacent to Mole National Park, Ghana (Photo by E.D. Wiafe)

- Conflict between herders and large predators such as cheetahs and leopards, resulting in loss of livestock and potential danger to human lives.
- Competition for water resources between wildlife and human communities during periods of drought, exacerbating tensions and increasing the risk of conflict.

These examples demonstrate the diverse nature of human–wildlife conflicts in Africa and the range of impacts they have on local communities. Addressing these conflicts requires a multi-faceted approach that integrates scientific research, community engagement, and policy interventions to ensure the coexistence of humans and wildlife while promoting sustainable development and conservation efforts. The Table 2.1

Table 2.1 List of Human–Wildlife Conflicts in Africa with mitigation measures

Case	Country	Cost/effect	Mitigation measures	Year of occurrence
Elephant crop raids	Kenya, Tanzania, Ghana	Destruction of farmlands, loss of livelihoods	Electric fencing, bee-hive deterrents, community patrols	2015 – Present
Lion attacks on livestock	Botswana, Kenya, Namibia	Loss of cattle, economic hardship for farmers	Predator-proof enclosures, compensation schemes, deterrents	2010 – Present
Hippo attacks on fishermen	Malawi, Zambia, Mozambique	Human injuries, deaths, loss of fishing equipment	Restricting fishing zones, buffer zones, warning systems	2012 – Present
Leopard attacks on livestock	South Africa, Zimbabwe	Economic losses for livestock farmers	Livestock guarding dogs, community awareness	2018 – Present
Crocodile attacks on humans	Uganda, Nigeria, Tanzania	Human deaths and injuries near water bodies	Safe water access points, fencing of riverbanks, awareness	2017 – Present
Bushmeat poaching conflicts	Cameroon, Central African Republic	Loss of endangered species, legal consequences	Alternative livelihoods, anti-poaching patrols, eco-tourism	2005 – Present
Hyena predation on livestock	Ethiopia, Sudan	Cattle, sheep, and goat losses affecting food security	Night enclosures, non-lethal deterrents	2016 – Present
Monkey crop raiding	Rwanda, Uganda, Ghana	Reduced farm productivity, loss of income	Scare tactics, planting non-palatable crops as buffers	2014 – Present
Human encroachment into wildlife habitats	Nigeria, DR Congo, Ivory Coast	Habitat loss, increased conflict, biodiversity decline	Land-use planning, protected area management, community-based conservation	2000 – Present
Buffalo aggression in rural areas	Tanzania, Zimbabwe, Zambia	Injuries and fatalities, destruction of property	Early warning systems, controlled grazing zones	2019 – Present
Snake bites in farming areas	Burkina Faso, South Africa, Kenya	Human fatalities, increased medical costs	Awareness campaigns, provision of anti-venom treatments	2013 – Present
Gorilla–human interactions in agricultural zones	DR Congo, Rwanda, Uganda	Crop destruction, potential disease transmission	Conservation corridors, eco-tourism incentives	2010 – Present
Zebra competition with livestock for grazing	Namibia, Kenya, Botswana	Overgrazing, food scarcity for domestic animals	Grazing management, rotational grazing techniques	2020 – Present
Wild dog predation on small livestock	Zambia, Botswana, South Africa	Loss of goats, sheep, and poultry	Conservation awareness, securing enclosures, compensation	2011 – Present

(*Source*: Compiled from International Union for Conservation of Nature (IUCN) Reports, World Wide Fund for Nature (WWF) Data, and peer-reviewed case studies on Human–Wildlife Conflicts in Africa, 2005–2023.)

provides a synthesis of case studies and mitigation measures on human-wildlife conflicts in Africa.

2.4 Mitigation and Coexistence Strategies

It is important to recognize that not all situations of human–wildlife conflict can be easily resolved, and in some instances, the only realistic goal may be in reducing the impact of the conflict on one of the parties involved. In situations where human safety is a concern, the most sensible course of action may be to translocate or euthanize the animal in question. When dealing with subsistence farmers living in poverty who are suffering agricultural damage, it may be unrealistic to implement measures aimed at conserving the wildlife if there are no short-term benefits provided to these people. This involved participatory workshops with local people and stakeholders to identify problems, causes, and possible solutions with the development of an action plan aimed at reducing adverse impacts on both wildlife and the local people.

Mitigation measures of human–wildlife conflict can be adopted to accomplish the goal of reducing or avoiding the conflict, which is negatively affecting both wildlife and human populations. Implementing the right mitigation action requires an understanding of the situation and the underlying causes of the conflict. Mitigation is an iterative process, whereby management is used to test hypotheses, and results are monitored to provide information to be used in refining the management actions. An adaptive management approach is highly recommended when dealing with human–wildlife conflict. This allows for changes to be made as results of mitigation actions are analyzed, and it is especially useful when working with diverse groups of stakeholders, each with different perceptions about the wildlife involved.

The first step involves evaluating the significance of the interaction between people and wildlife. Rural communities typically have a high dependency on natural resources, with agriculture being the main livelihood. Understanding that the need for food and income is universal, the type of land available for agriculture often coincides with prime wildlife habitat. Subsequently, both initiatives clash, creating a lose-lose situation whereby both parties incur costs. It has been suggested that wildlife often constitutes a form of land-use competition with agricultural communities and a "resource drain," often affecting the poorest members of society. This means that the provision of benefits for rural communities may directly decrease the levels of human–wildlife conflict as the costs of living with wildlife decrease. If there can be successful mitigation for HWC, it may mean that wildlife effectively subsidizes the development of rural communities. In cases involving government or privately owned protected areas, human encroachment and/or wildlife populations spilling out of reserves can still have cases of HWC. This is especially important to protect the livelihoods of people employed in the conservation sector, many of whom reside in rural communities. Wildlife providing benefits for communities is another reason to attempt mitigating HWC; by retaining the resources of agriculture and avoiding conflict situations, communities will still receive benefit from wildlife itself without incurring costs. A large emphasis on this comes in the form of damage control, and a crucial component to the economics of HWC is identifying critical areas which put the most costs on conservation and prevention of HWC around those areas. This often involves looking at high-cost low frequency events such as predator livestock depredation. Knowing that the broadest initiative in conservation is to increase the levels of awareness and education of HWC, identifying the opportunity costs between HWC and education will often conclude that prevention measures are a cost-effective way to avoid future losses of both wildlife and resources. Finally, for areas of HWC which are intensified through war conflict or with post-conflict societies, HWC can inhibit the potential for peace and independence, with pockets of communities being cut off from aid agencies and development programs. In these cases, strong mitigation measures can create a bridge for more positive community relations and aid in the healing process for both the people and war-torn wildlife.

Mitigating human–wildlife conflict in Africa is crucial for promoting coexistence between people and wildlife while minimizing negative impacts on both. Here are several mitigation measures commonly employed in Africa:

Physical Barriers: Constructing fences, trenches, or barriers to deter wildlife from entering human settlements or agricultural areas. Implementing electric fences or predator-proof enclosures around livestock enclosures or crop fields. Installing beehive fences as a natural deterrent for elephants, as they are averse to bees' stings (King et al. 2009).

Land-Use Planning: Designating wildlife corridors and buffer zones to allow for the free movement of animals between protected areas and to minimize conflicts with human activities. Encouraging sustainable land-use practices, such as agroforestry or rotational grazing, that maintain habitat connectivity while supporting local livelihoods (Naughton-Treves et al. 2003).

Community-Based Conservation: Engaging local communities in conservation efforts through participatory approaches, such as community conservancies or wildlife management committees. Providing incentives for communities to coexist with wildlife, such as revenue-sharing schemes from ecotourism or compensation for crop or livestock losses (Madden 2004).

Non-Lethal Deterrents: Implementing nonlethal deterrents, such as bright lights, noise makers, or pepper spray, to deter wildlife from approaching human settlements or livestock.

Using trained guard animals, such as livestock guardian dogs or donkeys, to protect livestock from predation (Gehring et al. 2010).

Education and Awareness: Conducting educational outreach programs to raise awareness about human–wildlife conflict and promote strategies for coexistence.

Providing training and capacity-building opportunities for local communities in conflict mitigation techniques and sustainable natural resource management practices.

Implementing a combination of these mitigation measures tailored to specific contexts and species can help reduce human–wildlife conflict and promote harmonious relationships between people and wildlife in Africa (Frank et al. 2020).

References

Conover MR (1998) Perceptions of American agricultural producers about Wildlife on their farms and ranches. Wildl Soc Bull 26(3):597–604. https://www.jstor.org

Conover MR (2001). Resolving wildlife conflicts: the science of wildlife damage management. CRC Press. https://doi.org/10.1201/9781420032581

Frank B, Glikman JA, Marchini S, Gompper ME (2020) Human dimensions of large carnivore conservation and coexistence. Conserv Biol 34(1):1–10

Gehring TM, VerCauteren KC, Landry JM (2010) Livestock protection dogs in the 21st century: is an ancient tool relevant to modern conservation challenges? BioScience 60(4):299–308

International Union for Conservation of Nature (IUCN), World Wide Fund for Nature (WWF), & Various Authors (2005–2023) Human–wildlife conflicts and mitigation strategies across Africa: a synthesis of case studies and field reports. Retrieved from https://www.iucn.org and https://www.worldwildlife.org

King LE, Lawrence A, Douglas-Hamilton I, Vollrath F, Sitati N (2009) Beehive fences as a sustainable solution to human-elephant conflict: proceedings of the 2007 Rufford Small Grants Conference. Conservation of Large Mammals, pp 77–79

Madden F (2004) Creating coexistence between humans and wildlife: global perspectives on local efforts to address human–wildlife conflict. Hum Dimens Wildl 9(4):247–257

Muhumuza M, Balkwill K (2013) Factors affecting the success of conserving biodiversity in national parks: a review of case studies from Africa. Int J Biodiv 2013. Article 798101. https://doi.org/10.1155/2013/798101

Naughton-Treves L, Grossberg R, Treves A (2003) Paying for tolerance: rural citizens' attitudes toward wolf depredation and compensation. Conserv Biol 17(6):1500–1511

3 Agriculture and Food Security

3.1 The Agricultural Landscape in Africa

The chapter considers many themes that are often interlinked. The essays that relate to ecological imperialism and tropical deforestation give an insight into the damage of foreign influence on African landscapes. This can be contrasted with the examination of pre-colonial and colonial agricultural systems. The diffusion of new world crops in Africa is considered by Robbins and Faursby, bear in mind that one of the most influential factors of this change was the sheer extent of destruction caused in the native landscape by colonial powers. The recent changes in landscape and the cause of them are an important theme and a number of articles give an insight into the highly influential structural adjustment programs and the course of independence in Africa. Overall, there is a message that change is not always a process of progression and modernization has had mixed results from the different regions of Africa. The past four decades have been an era of rapid globalization and change in the global economy has had a significant impact on African agricultural systems, an example can be found in the recent studies of agribusiness in Africa (For Translation and Literary Studies & Hamad 2021; Ahn et al. 2022; Assistant and Gaiya 2023).

The agricultural landscape in Africa is a study of the impact human societies have had on the natural African landscape. The landscape has united humanity and nature in the continuous pursuit of survival. The richness of the land and the diversity of resources found in Africa were causal to the long-term interaction between humankind and their environment. It is difficult to speak of a single African landscape, and analyse the various landscapes and the transitions occurring within them. The great diversity of physical landforms and the ever-changing climatic conditions have necessitated adaptation of new farming systems and a constant change in agricultural practices. The results reflect the premise that the agricultural landscape is an impression of both the culture and the environment and the two are inextricably linked. This collection of articles provides evidence of the many changes that have occurred in African agricultural landscapes through time and a comparison of the different processes of change in different regions. Economic and political factors are perhaps now the most influential they have ever been in shaping the African landscape, thus some of the essays have a modern context and give insight into the implications of current

E. D. Wiafe, *Wildlife Conservation and Agriculture Development in Africa*, Sustainable Development Goals Series, https://doi.org/10.1007/978-3-032-09233-5_3

agricultural trends (Dasgupta and Johnson-Kanu 2020; Gamo Longang et al. 2023; Sackey Teye and Tetteh Quarshie 2021).

3.2 Historical Overview and Evolution of Agriculture in Africa

3.2.1 Precolonial Agriculture

The existence of a complex and highly productive precolonial agricultural system in Africa stands as a pivotal and profound theme that enables us to truly grasp the rich agrarian history of this diverse continent. Across various regions of Africa, such as the vast savannas, the lush and humid highlands, and the temperate uplands, the precolonial era witnessed remarkable development and cultivation practices that were of such magnitude that they resulted in extensive deforestation and substantial soil erosion. It is crucial to dispel any notion that sub-Saharan Africa solely engaged in mobile gardening during this period, as this misperception fails to capture the profound diversity of agricultural practices that indeed existed (Achieng et al. 2023; Asante Gyamerah and Ikpe 2021; Dasgupta and Johnson-Kanu 2020; Seranne and Anka 2005). While it is certainly accurate that historical Bantu-speaking communities, which comprise the vast majority of the present inhabitants in this region, embraced a system of shifting cultivation specifically designed to sustain the fertility of the soil, it is essential to recognize that numerous African societies, particularly in areas like present-day Cameroon and Nigeria, engaged in a wide array of intensive and continuous cultivation methods. The African landscape witnessed a tapestry of agricultural varieties that were shaped by the distinctive environments in which they thrived, as well as the historical movements of various peoples through migrations, trade networks, conflicts, and the desperate flight from the perils of slave raiding and slave trading. The latter, in particular, exerted a profound influence on the fabric of societies in certain regions, leaving an indelible mark on their agricultural practices (Ricome et al. 2023; Yin et al. 2023; Seidensticker and V. M. Jungnickel 2023; Walther and Retaille 2019; Castellanza 2019; Polanco and E. J.Newman 2023). For instance, scholarly research suggests that during the early nineteenth century, agricultural communities residing in the region surrounding the Great Lakes of East Africa were compelled to abandon their ancestral lands and seek refuge on the islands due to the pervasive activities of slave raiders. This disruptive force instigated a profound shift in the dynamics of agricultural practices as communities were forced to adapt and relocate in order to ensure their survival and maintain their social fabric. The resilience and ingenuity displayed by these communities in the face of such external pressures serve as a testament to the undeniable complexity and diversity that existed within African agriculture during the precolonial era.

3.2.2 Colonial Influence on Agricultural Practices and Methods

The period of colonial contact also had a profound and far-reaching impact upon precolonial agrarian systems in Africa, which has been a continuous and ever-evolving source of agrarian change throughout history. The purposes and intentions behind colonial administration varied considerably, depending upon the nation by which the colony was ruled, thereby shaping the course of events in distinct ways. There were four primary strategies that were employed in the pursuit of colonial dominance and control:

- Christian missionary activity, which was primarily championed by the British;
- the establishment and development of settler colonies, primarily undertaken by the French for Africa;
- the exploitative utilization of African natural and human resources as a means to dismantle the cruel and vile institution of slavery; and lastly,
- the creation and fostering of designated African colonies with the primary objective of serving the exclusive interests and requirements of the colonizing nation, with the Portuguese taking the lead in this aspect.

(Dasgupta and Johnson-Kanu 2020; Neupert-Wentz and Müller-Crepon 2021; Asante Gyamerah and Ikpe 2021; Dasgupta and Johnson-Kanu 2020; Spindelman and Arturo Crouch 2024).

Every method of colonization wielded its own distinct and unique effects upon the agrarian systems prevalent at the time, but amidst the diversities and complexities, there are certain fundamental similarities that can be observed in the transformative changes that unfolded and persisted throughout the colonial period and well into the post-colonial era. The lasting impacts of colonialism upon agrarian systems in Africa cannot be understated, as they have played an indelible role in shaping the course of agricultural practices and the very essence of society in this region. By examining and understanding the multifaceted ways in which colonialism intersected with agriculture, a deeper comprehension of the evolving dynamics between humans and the land can be attained, enabling a more nuanced perspective on the intricacies of agrarian change in Africa.

(Dasgupta and Johnson-Kanu 2020; Dasgupta and Johnson-Kanu 2020; Dasgupta and Johnson-Kanu 2020; Spindelman and Arturo Crouch 2024; Dasgupta and Johnson-Kanu 2020; Asante Gyamerah and Ikpe 2021; Dasgupta and Johnson-Kanu 2020; Neupert Wentz and Müller-Crepon 2021; Dasgupta and Johnson-Kanu 2020; Spindelman and Arturo Crouch 2024).

The introduction of colonial authority saw changes in land holding regimes. Precolonial land tenure systems were diverse, involving a mix of communal and individual land ownership. The division of land according to ethnic group was common, with each group exercising rights over a specific area of territory. The way rights to land were held, used and transferred and the effectiveness with which they were enforced varied considerably. Colonial economies were market-oriented and competitive, and the colonizers sought to exploit African land and resources to grow crops with high economic value. This brought African land into the world economy, with a resultant restructuring of agrarian systems. In West Africa, the establishment of cash crop production, such as cocoa and groundnuts, saw the application of intensive methods of cultivation and in some cases a change in the use of land from traditional forms of shifting cultivation or pastoralism to more permanent farming (Dasgupta and Johnson-Kanu 2020; Dasgupta and Johnson-Kanu 2020; Dasgupta and Johnson-Kanu 2020; Spindelman and Arturo Crouch 2024; Dasgupta and Johnson-Kanu 2020; Asante Gyamerah and Ikpe 2021; Dasgupta and Johnson-Kanu 2020; Neupert-Wentz and Müller-Crepon 2021).

3.2.3 Post-Independence Agriculture

In the post-independence period, governments in Africa continued to give low priority to agriculture. They were able to generate revenues from other sources, such as copper, oil, and foreign aid, and thus were not dependent on food production. Therefore, during the 1960s and 1970s, African countries were net importers of food and were able to get by without major improvements in their agricultural sectors. This led to unsustainable food import bills that many African nations are still grappling with today. The gain of political independence in African countries also led to a shift in power and priorities of the ruling class. Bearing in mind the influence still evident from the former colonial governments, many post-independence leaders began to shift toward cash crop production, particularly products that had been exported to the former colonial rulers. This was due to forced marketing boards and commodity pricing, and most newly independent African nations found themselves in a situation where they were discouraged from developing any form of import substitution industrialization as a means of getting out of the colonial legacy of trade and dependency on other countries. However, it should be noted that this lack of emphasis on agriculture came with severe consequences for the African population. The neglect of the agricultural sector resulted in limited job opportunities and income generation for rural communities, exacerbating issues of poverty, inequality, and food insecurity. Additionally, the heavy reliance on cash crops for export left African nations vulnerable to fluctuations in global market prices, making them susceptible to economic instability. Furthermore, the focus on cash crops often meant that fertile land was solely utilized for commercial purposes, leading to environmental degradation and the loss of biodiversity. These detrimental effects have had long-lasting repercussions on the African continent, hindering sustainable development and perpetuating a cycle of dependency on foreign aid and food imports. In recent years, there has been a growing recognition of the importance of revitalizing the agricultural sector in Africa. Governments and international organizations are now prioritizing agricultural investments, promoting technological advancements, and supporting small-scale farmers to enhance productivity, income, and food security. Efforts are being made to adopt sustainable farming practices, improve irrigation systems, and enhance value chains to ensure that agriculture becomes a key driver of inclusive economic growth and poverty reduction. By empowering local farmers and fostering agricultural innovation, Africa has the potential to not only meet its own food needs but also become a major player in global food markets. It is crucial that the lessons learned from past mistakes are taken into account and policies are implemented to address the structural challenges that have hindered the agricultural sector's development. Through sustainable and inclusive agricultural practices, Africa can secure a prosperous and self-sufficient future, harnessing the immense potential of its vast natural resources for the benefit of its people and the continent as a whole (Michler et al. 2020; Enisan Akinlo and Michael Ojo 2022; Kruseman 2022; Michler et al. 2020; Nivievskyi et al. 2023; Dasgupta and Johnson-Kanu 2020; Enisan Akinlo and Michael Ojo 2022; Christophe Jacques and Defourny 2019; Djihad et al. 2021; Asante Gyamerah and Ikpe 2021).

3.3 Agricultural Practices

Many different and sometimes completely opposing views have been expressed on the nature and value of African farming systems. This is not surprising given the great diversity found, and has also produced some overlap. However, a convenient, although artificial dichotomy, can be drawn between farming systems which are primarily aimed at producing cash crops for an external market to earn income for the farmers, and those which are primarily aimed at producing food for subsistence. For better or worse, agriculture is being increasingly oriented towards cash crops, since much of the population is already dependent upon the money economy for survival, and better returns can often be had compared to food crops. The expansion of cash cropping has frequently been associated with dispossession of peasants from their lands and establishment of large-scale commercial farming. At one extreme is the vast monocultured plantations for tea, coffee, sugar, oil palm, soybeans etc., often owned and run by multinational companies, which occur in some countries. In these cases, the normal rural landscape is destroyed, local people evicted, and primal forest or savannah of high biological diversity replaces a huge area which has only one crop species and hence 1/1 diversity! At the other end are the smallholdings which produce a cash crop with the aim of supplementing food supplies or to provide cash where no other employment is available. These smallholdings, often owned by independent farmers, play a vital role in local economies by contributing to the overall food security and income generation in their communities. In addition, they also serve as important centers of agricultural knowledge and experimentation, as farmers constantly adapt their practices to changing environmental conditions and market demands. Furthermore, it is worth noting that the expansion of cash cropping does not occur in isolation. It is influenced by a myriad of factors such as government policies, climate change, technological advancements, and international trade agreements. All these factors shape the landscape of African farming systems and contribute to the complex web of relationships between farmers, markets, and ecosystems. It is important to acknowledge and respect the diversity of African farming systems and the various reasons why farmers choose to cultivate cash crops or focus on subsistence farming. Any discussion or approach towards agricultural development in Africa should take into account the socio-economic, cultural, and environmental contexts in order to ensure sustainable and inclusive growth. By embracing this holistic perspective, it becomes possible to support the unique needs and aspirations of African farmers while also promoting resilience, equity, and environmental conservation in the face of an uncertain future.

3.3.1 Traditional Farming Methods

African traditional agriculture, which has been expertly practiced for countless centuries, heavily relies on labor-intensive manual techniques and a careful approach to shifting cultivation practices in order to meticulously ensure the sustained well-being of local communities. Unlike modern agricultural practices, which unfortunately can have severe detrimental effects on the environment, these time-honored traditional methods offer an inherently more sustainable and environmentally conscious approach to farming. Shifting cultivation, a fundamental and cherished component of African traditional farming, involves a meticulously orchestrated process consisting of the meticulous clearing of woodland, the controlled burning of scrub, and the attentive tending to crops until they reach full maturity and are ready to be harvested, all while seamlessly respecting the natural cycles of the land. Once the land has been thoughtfully utilized and all its potential harnessed, it is left undisturbed to peacefully regenerate naturally, rejuvenating its vital resources as it quietly replenishes itself before the next cultivation cycle begins anew. Through a delicate balance of human intervention and passive observation, African traditional agriculture cultivates not just crops, but also a profound connection between humans and their surrounding environment. While, regrettably, there have been instances in the past where the use of shifting cultivation has been unjustly prohibited, there is now an encouraging and imminent shift in perspective, with a growing awareness and a necessary reconsideration of its unparalleled potential benefits. As passionate experts and dedicated communities ardently study the long-term impacts of traditional agriculture, it becomes increasingly evident that these time-tested age-old practices foster an unparalleled harmonic interdependence between humans and their surrounding environment, truly epitomizing the remarkable wisdom, ingenuity, and resilience of African agricultural traditions that have stood the test of time. This interconnectedness permeates every aspect of traditional agricultural practices, from the careful selection of crop varieties to the meticulous observation of weather patterns and natural phenomena. Every decision made by African traditional farmers is rooted in a deep understanding of their environment, gained through generations of accumulated knowledge and experience. Furthermore, the sustainability of African traditional agriculture extends beyond the cultivation of crops. It encompasses the entire ecosystem, recognizing the importance of healthy soil, clean water sources, and diverse wildlife habitats. These practices actively preserve biodiversity and contribute to the overall health and well-being of the land. In addition to the environmental benefits, traditional agriculture also plays a vital role in the social fabric of African communities. It fosters a sense of unity, cooperation, and shared responsibility among community members. The process of land preparation, crop cultivation, and harvest becomes a communal endeavor, strengthening bonds and cultural traditions. As Africa embraces its rich heritage and explores ways to meet the evolving challenges of agriculture, the importance of traditional practices stands out as an invaluable resource. By combining the wisdom of the past with modern innovations and scientific advancements, African traditional agriculture has the potential to offer sustainable solutions to global food security, climate change mitigation, and the preservation of cultural heritage. In conclusion, African traditional agriculture stands as a testament to the harmonious relationship between humans and the environment. It showcases the resilience, ingenuity, and deep-rooted wisdom that have guided generations of farmers in nurturing their land and sustaining their communities. Through the careful cultivation of crops and the preservation of natural resources, African traditional agriculture exemplifies a sustainable path forward, inspiring a new generation of farmers and researchers to learn from the past and shape a brighter future for all (Hasan et al. 2020; Raimi et al. 2021; Akanmu et al. 2023; Nemade et al. 2023; Ema et al. 2023; Ogura et al. 2024; Jayasekara 2021; Kansanga et al. 2020; Alajangi et al. 2021; Rai et al. 2023).

Modern Farming Techniques have greatly revolutionized agricultural practices. With the help of innovative technologies and scientific advancements, farmers now have access to a wide range of tools and methods that enhance productivity and sustainability. These techniques encompass various aspects of farming, including irrigation systems, mechanization, genetic engineering, and precision agriculture. Irrigation systems play a vital role in modern farming (Fig. 3.1). Advanced techniques such as drip irrigation, sprinkler systems, and hydroponics ensure optimal water usage, preventing wastage and maximizing crop yields. This not only conserves water resources but also allows farmers to cultivate crops in regions that were once considered unsuitable for agriculture. Mechanization has significantly reduced manual labor on farms. From tractors and combine harvesters to automated seeders and feeders, machinery has revolutionized the way farmers work. It increases efficiency, saves time, and enables farmers to manage larger areas of land without exhausting their physical resources. Genetic engineering has enabled the development of genetically modified organisms (GMOs) that offer increased resistance to pests, diseases, and adverse environmental conditions. These genetically enhanced crops can produce higher yields with improved nutrition levels. They also contribute to more sustainable farming by reducing the need for chemical pesticides and herbicides. Precision agriculture, also known as satellite farming, utilizes remote sensing technologies and data analysis to allow farmers to make informed decisions about resource allocation. By monitoring aspects like soil moisture, nutrient levels, and crop health, farmers can optimize inputs like fertilizers and pesticides, minimizing waste and environmental impact. Furthermore, modern

Fig. 3.1 Irrigation system supports all season farming (Photo by E.A. Tetteh)

farming techniques encourage organic farming practices and sustainable agriculture. Integrated pest management, crop rotation, agroforestry, and composting are all embraced to reduce reliance on synthetic inputs and promote a healthier, more environmentally friendly approach to farming. In conclusion, modern farming techniques have revolutionized the agricultural industry by improving efficiency, productivity, and sustainability. These methods harness the power of technology and scientific innovation to optimize resource usage and produce higher-quality crops. By adopting these practices, farmers can meet the growing demands for food in a responsible and environmentally conscious manner. The adoption of these techniques is crucial in ensuring food security for the rising global population and achieving a sustainable future for agriculture. It is through the continuous exploration of cutting-edge technologies and the dissemination of knowledge and best practices that we can realize the full potential of modern farming techniques. Through collaboration between researchers, farmers, policymakers, and industry experts, we can further refine and expand these techniques to face the challenges posed by climate change and limited resources. By embracing innovation and sustainability, we can create a future where farming is not only productive but also environmentally friendly, ensuring the well-being of both current and future generations (Amusan and Oyewole 2023; Fastellini and Schillaci 2020; Shaikh et al. 2022; Karunathilake et al. 2023; Raj et al. 2021; Duncan et al. 2021; Gokool et al. 2023; Kwaghtyo and Eke 2023; Udoumoh et al. 2021).

3.3.2 Sustainable Agriculture

The basis for the Indigenous Soil Fertility Initiative lies in the international acceptance of the processes of Participatory Learning Action Research (PLAR) and Legume Best Bets, and in the experience of the people who have been involved in soil fertility replenishment (SFR) activities over a number of years. Treatments, best bets, and management guidelines will be based on the synthesis of the practical experiences and observations of a wide range of development workers and farmers, including those involved in related exercises in other countries, linked with experimentation to fill knowledge gaps. Furthermore, the use of a variety of on and off-station PLAR methods, in conjunction with the work on legumes and rotations, offers an important opportunity to test options and innovations in improved SFR and to understand farmers' decision-making in response to problems and opportunities for managing resources and investments in complex farming systems. This integrated approach ensures that the Indigenous Soil Fertility Initiative remains a comprehensive and holistic endeavor, encompassing various strategies and perspectives on soil fertility replenishment. By gathering insights and data from both local communities and international sources, the initiative can build a robust framework that considers the unique challenges and opportunities faced by farmers in different regions. Through the collaboration between development workers, farmers, and researchers, innovative solutions and practical guidelines can be developed to address specific concerns related to soil fertility. Notably, the Indigenous Soil Fertility Initiative recognizes the importance of sustainable practices and the need to strike a balance between productivity and environmental stewardship. With this in mind, the initiative aims to promote methods that not only improve soil fertility but also minimize negative impacts on the ecosystem. By taking into account the principles of agroecology and regenerative agriculture, the initiative seeks to empower farmers with knowledge and tools that enable them to cultivate healthy and resilient farming systems. Overall, the Indigenous Soil Fertility Initiative signifies a significant step forward in the field of sustainable agricultural development. Through its comprehensive approach, this initiative strives to enhance the livelihoods of farmers, protect the environment, and contribute to the overall well-being of communities.

3.4 Major Crops Cultivated in Africa

The diversity of crop types grown in Africa is incredibly vast and extensive. Amongst the array of cereals, it is remarkable to note that sorghum and millet have been especially adapted to thrive in the challenging conditions of the semi-arid tropics, solidifying them as traditional crops deeply rooted in African agriculture. However, it is fascinating to observe how maize, wheat, and rice, although comparatively recent introductions to the continent, have achieved remarkable growth and cultivation, surpassing even the traditional cereals in terms of land area devoted to their cultivation. Furthermore, it is worth mentioning that Africa serves as the birthplace and center of origin for an abundance of cereals, such as finger millet and African rice, boasting an extensive genetic diversity that contributes to their overall resilience and adaptability. Moving beyond cereals, root crops play a crucial role in African agriculture, with cassava and sweet potato standing out as prime examples of traditional crops that serve as staples in various regions of Africa. Cassava holds particular importance in multiple parts of West Africa, where it acts as the primary source of carbohydrates, essentially becoming the lifeblood of these areas. Fascinatingly, even peanuts and sunflower, categorized as oilseeds, possess the unique distinction of also being considered "pulses" from an agronomic standpoint, owing to their notable oil content. Consequently, this alternative classification has proven to be remarkably advantageous and practical. Furthermore, soybean, an incredibly vital pulse, although not originally a traditional crop, has successfully adapted to diverse African conditions and showcases immense potential for exportation. In addition to these remarkable crop varieties, tuber and plantation crops play a pivotal role in sustaining and supporting countless small landholders throughout the continent. Among these crops, cocoa reigns supreme (Fig. 3.5), with its extensive cultivation. However, despite its prevalence, there are significant challenges that need to be addressed. Widespread diseases and low yields pose major obstacles to maximizing cocoa production and achieving its full potential. Nonetheless, with continued efforts and advancements in agricultural practices, there is great hope for overcoming these challenges and discovering new opportunities for growth and improvement within the African agricultural sector.

3.4.1 Cassava

Cassava (manioc, yuca) is a "stick crop" as it reproduces itself from cuttings (Fig. 3.2). A versatile and staple crop, most of the cassava produced in Africa is used for home consumption, with a relatively low proportion traded or sold on the market. This crop has traditionally been viewed as a food security crop, a rebellion crop, and a "women's crop". It has not been widely researched by agricultural scientists and social scientists despite its importance for local, regional, and national economies and its cultural significance. Cassava is labeled a "1000 day crop" as it can be left in the ground for long periods without losing its edibility. An allocation of labor to cassava effectively means that it provides "food insurance" for the family in that as a perennial (between 1 and 3 years depending on local disease), it can always be left in the ground until needed. The shift from shifting cultivation to semi-permanent or permanent cropping systems has had considerable influence on the allocation of land and labor to cassava. Since the 1960s, there has been a regional trend of the expansion of cassava cultivation. This is largely because cassava is now being grown in far more diverse agroecological zones than that of maize. In the past, cassava and maize cultivation were often sympatric, and cassava was labeled a "disease" or "famine" crop to be grown in marginal areas in case of crop failure. However, due to its versatility and increasing demand in various markets, cassava is now perceived as a profitable cash crop. Consequently, there is an influx of new labor being allocated to improve cassava agronomy and maximize its economic potential. This expansion has led to complex consequences on livelihoods and gender relations, challenging the traditional view of cassava as exclusively a "woman's crop." Interestingly, it is often men who spearhead the expansion of cassava into new areas as they recognize the lucrative market opportunities associated with it, seeking to generate cash income for their families and communities. As cassava continues to spread to new regions and attract attention from farmers, there is a growing need for comprehensive research and exploration into its agricultural

Fig. 3.2 Cassava (*Manihot utilissima*) farm (Photo by E. A. Tetteh)

and socio-economic implications. Despite being a long-established crop, cassava remains under-represented in scientific research, which limits our understanding of its full potential. By delving deeper into cassava's benefits, challenges, and complexities, scientists and agricultural experts can unlock strategies to harness its economic value sustainably. One particular aspect that has contributed to the expansion of cassava cultivation is the transition from traditional shifting cultivation practices to more permanent or semi-permanent cropping systems. This shift has significantly affected the allocation of land and labor dedicated to cassava production. Since the 1960s, there has been a noticeable regional trend towards the wider cultivation of cassava. Unlike maize, which is limited to specific agroecological zones, cassava has adapted to a broader range of environments. Previously, cassava and maize were often grown together, with cassava serving as a contingency crop for adverse conditions or crop failures. However, with its versatility and increasing demand across various markets, cassava has transformed into a profitable cash crop. Consequently, this transformation has attracted a surplus of labor and resources aimed at improving cassava agronomy and maximizing its economic potential. The expansion of cassava cultivation has had intricate ramifications for livelihoods and gender dynamics, challenging the conventional perception of cassava as solely a "woman's crop." Intriguingly, it is often men who drive the expansion of cassava into new territories. Recognizing the lucrative market opportunities associated with cassava, they aspire to generate cash income for their families and communities by capitalizing on its growth. The unprecedented growth and commercialization of cassava highlight the need to explore its impact comprehensively. Stakeholders must conduct in-depth research to grasp the multifaceted consequences this expansion brings to different societies. Examining the agronomic, economic, and social dimensions can shed light on how cassava can yield sustainable development, enhance food security, and promote gender equality. By recognizing the agency of men in promoting cassava expansion and harnessing the expertise of both male and female farmers, we can ensure a more inclusive and prosperous future for cassava cultivation (Kaluba 2021; Inegbedion et al. 2020; Ceballos et al. 2020; Scott 2021; Adebayo and Silberberger 2020; Adebayo 2023; Odekina 2023; Ikuemonisan et al. 2020; Cock and Connor 2021).

3.4.2 Sorghum

Sorghum, also known as durra, milo, jowar, or kaoliang, is a highly versatile and remarkably resilient crop that has been cultivated for centuries. It serves as a valuable staple food for millions of people all around the globe, and its popularity continues to soar due to its myriad of health benefits (Fig. 3.3). Sorghum boasts an astonishing array of vital vitamins and minerals, such as iron, magnesium, phosphorus, and an abundance of B vitamins that are essential for the body's optimal functioning. What's more, sorghum possesses the remarkable distinction of being completely gluten-free, thereby serving as an exceptional alternative for individuals suffering from gluten sensitivities or celiac disease. Its adaptable nature allows it to be seamlessly incorporated into a multitude of dishes, ranging from hearty soups and delectable stews to delectable loaves of bread and luscious desserts like sorghum syrup. The exceptional versatility and substantial nutritional value of sorghum make it an impeccable choice for adding delightful diversity to your meals. Whether you're seeking a captivating grain to introduce into your culinary repertoire or simply yearning for a remarkable gastronomic adventure, sorghum stands as a truly remarkable option that

Fig. 3.3 Farm of Sorghum (*Sorghum bicolor*) in Burkina Faso

is bound to tantalize your taste buds. Delve into the captivating world of sorghum, and unlock the culinary treasures that await with every captivating bite. Embark on a remarkable gastronomic journey that will leave you utterly captivated by all the delicious possibilities that this extraordinary grain has to offer!

3.4.3 Maize

Maize (*Zea mays*), also known as corn, is a cereal grain that originated in Mesoamerica thousands of years ago. Today, it is one of the most widely cultivated and consumed crops in the world, valued for its versatility, nutritional value, and cultural significance. Maize has played a vital role in the development of civilizations, serving as a staple food in many cultures and cuisines globally. Its cultivation spread across the Americas and eventually reached other parts of the world through colonial expeditions and trade. The various uses of maize have contributed to its popularity and widespread cultivation (Fig. 3.4). One of the primary uses of maize is as a food source for both humans and animals. The grains can be consumed directly as corn on the cob or processed into a variety of products that have become essential components of many cuisines. Cornmeal, corn flour, and corn oil are just a few examples of the diverse products derived from maize. These ingredients are commonly used in bread, tortillas, cereal, snacks, and countless other dishes. In addition to its culinary uses, maize is also a crucial feed for livestock. Its high nutritional content and availability make it an ideal choice for animals, contributing to the growth and productivity of livestock around the world. The vibrant colors of maize, such as yellow, white, or multicolored kernels, add an appealing visual element to both traditional and modern recipes. From vibrant salsas to hearty stews, maize has the power to transform any dish into a feast for the senses. Throughout history, the cultivation and consumption of maize have evolved with advancements in agricultural techniques and scientific research. Farmers continue to innovate, striving to improve yield, disease resistance, and nutritional value. These efforts have led to the development of hybrid varieties and the implementation of sustainable farming practices, ensuring that maize remains a reliable and resilient crop. Maize holds a special place in many cultures and traditions worldwide. From ancient Mesoamerican civilizations to modern-day farmers, its significance is deeply rooted. Festivals, rituals, and celebrations often revolve around maize, marking its importance in society. Whether enjoyed as a simple roasted corn on the cob, fluffy popcorn, or warm and soft tortillas, maize brings comfort and nourishment to people worldwide. Its rich history and versatility make it an integral part of global culinary traditions. As we look to the future, the continued cultivation and consumption of maize will undoubtedly play a significant role in sustaining and nourishing communities for centuries to come.

3.4.4 Millet

Early in this century, it was widely recognized that the Sudanic, Guinean, and Northern regions of Nigeria encompassed a rich tapestry of well-established field agriculture systems that stretched across the expansive open savanna zones. The fundamental cornerstone of these intricate systems, discernible through meticulous botanical analysis, was found to be none other than the venerable millet plant in the Sahelian areas and sorghum in the southern and middle belt areas. Delving deeper into this immensely captivating and endlessly intriguing realm, we unveil the wondrous and enchanting world of the millets, a remarkably diverse and vast group of small-seeded species, comprising an astonishing array of highly sought-after cereal crops and grains. These extraordinary plants, with their unparalleled beauty and unbelievable resilience, have achieved unparalleled worldwide recognition and unequivocal acclaim, thriving in widespread cultivation for their immense value in both sustenance and fodder production. Though they may not fall under a single, easily categorizable taxonomic group, their collective functional unity is palpable, forged by their extraordinary and remarkably similar characteristics that unite them as a forceful and steadfast

Fig. 3.4 A young maize (*Zea mays*) farm (**A**) and Matured maize farm (**B**)

entity. Unsurprisingly, the millets have emerged as an indispensable and vital crop, a veritable lifeline, in the semi-arid tropics of mesmerizing Asia and breathtaking Africa, where their abundantly fruitful harvests grace nations such as the awe-inspiring India, the magnificent Mali, the resplendent Nigeria, and the awe-inspiring Niger. These nations, among many others, have undeniably reaped abundant and resplendent benefits from the millets' bountiful harvests, cultivating a symbiotic relationship with these exceptional plants that enriches their societies and strengthens their agricultural prowess. In fact, it shines as an illuminating beacon that developed countries, with their boundless knowledge and unyielding dedication, have earnestly embarked upon transformative and visionary programs, tasked with the noble mission of introducing these extraordinary and unparalleled crops to a broader and more diverse consumer base. In recognizing the millets' immense potential and the invaluable contributions they can make to global food security, these remarkable programs serve as testament to the resounding impact and unprecedented influence of these exceptional plants. The millets, with their awe-inspiring versatility, captivating nutritional profiles, and unparalleled adaptability, hold the promise of reshaping the future of humanity's sustenance as we know it. As the world grapples with the ever-pressing challenges of population growth, climate change, and the depletion of natural resources, the significance of the millets becomes even more pronounced and their effect even more profound. Thus, as we journey deeper into the mesmerizing realm of these extraordinary plants, we cannot help but marvel at their incredible essence and transformative capabilities. They stand as a testament to nature's unrivaled ingenuity and persistent determination to thrive against all odds. The millets, with their undeniable allure and limitless potential, beckon us to explore, appreciate, and embrace their bountiful offerings, leading us towards a future of unparalleled abundance and nourishment, where global food security is not merely a dream, but an unwavering reality. One major cash crop cultivated in rainforest of Africa which is worth mentioning is Cocoa (Fig. 3.5) (Table 3.1) (Cock and Connor 2021; Ikuemonisan et al. 2020; Odekina 2023; Adebayo 2023; Adebayo and Silberberger 2020; Scott 2021; Ceballos et al. 2020; Inegbedion et al. 2020; Kaluba 2021).

Fig. 3.5 A picture of cocoa (*Theobroma cacao*) farm in African rainforest zone (photo by E.D. Wiafe)

Table 3.1 Major crops cultivated in Africa, the nutritional and economic values, climatic and cultivation methods

Name of crop (Scientific Name)	Country	Nutritional value	Economic value	Climatic suitability	Cultivation methods
Maize (*Zea mays*)	Nigeria, South Africa, Kenya, Ghana	Rich in carbohydrates, dietary fiber, and some proteins	Staple food, animal feed, biofuel production	Warm temperate to tropical, 20–30°C, well-drained soils	Direct seeding, requires moderate rainfall
Cassava (Manihot esculenta)	Nigeria, DR Congo, Ghana, Tanzania	High in carbohydrates, provides energy, gluten-free	Used in flour, starch, bioethanol production	Tropical, 25–30°C, drought-tolerant	Planted from stem cuttings, grows in poor soils
Rice (*Oryza sativa*)	Egypt, Nigeria, Madagascar, Senegal	Source of carbohydrates, contains B vitamins	High market demand, export crop, food security	Warm, humid areas, 20–35°C, requires abundant water	Transplanting/ direct seeding, flooded paddies
Sorghum (*Sorghum bicolor*)	Ethiopia, Sudan, Burkina Faso, Niger	Rich in fiber, antioxidants, gluten-free	Drought-resistant, used for brewing, animal feed	Semi-arid, 25–35°C, thrives in low rainfall areas	Direct seeding, drought-tolerant
Millet (*Pennisetum glaucum*)	Mali, Chad, Niger, Sudan	High in protein, iron, and fiber	Resilient in dry climates, staple food, beer production	Hot, arid climates, 25–35°C, drought-resistant	Broadcast seeding, low water requirements
Wheat (*Triticum aestivum*)	Egypt, Ethiopia, Sudan, South Africa	High in fiber, protein, B vitamins	Bread, pasta, high export value	Cool to temperate, 15–25°C, needs well-drained soils	Direct seeding, irrigation in dry areas
Yam (*Dioscorea* spp.)	Nigeria, Ghana, Benin, Ivory Coast	High in carbohydrates, vitamin C, potassium	Staple food, processed into flour, traditional dishes	Humid tropics, 25–30°C, well-drained soils	Planted from tuber cuttings, needs staking
Plantain (*Musa paradisiaca*)	Cameroon, Ghana, Uganda, DR Congo	Good source of potassium, fiber, and vitamins A & C	Consumed as food, used in chips and flour production	Tropical, 25–35°C, high humidity, rich soils	Suckers propagation, requires mulching
Cocoa (*Theobroma cacao*)	Ivory Coast, Ghana, Nigeria, Cameroon	Antioxidants, iron, magnesium	Major export crop, used in chocolate production	Humid tropics, 22–28°C, high rainfall	Planted from seeds or grafting, shade required
Coffee (*Coffea arabica, C. robusta*)	Ethiopia, Kenya, Uganda, Rwanda	Contains caffeine, antioxidants	High export earnings, widely traded commodity	Cool highlands, 15–24°C, requires well-drained soils	Grown from seeds, requires pruning
Tea (*Camellia sinensis*)	Kenya, Malawi, Rwanda, Zimbabwe	Antioxidants, boosts metabolism	Major export crop, supports employment in tea estates	Cool, misty highlands, 15–25°C, acidic soils	Grown from cuttings, requires frequent pruning
Groundnuts (*Arachis hypogaea*)	Senegal, Nigeria, Ghana, Sudan	High in protein, healthy fats, and vitamins	Used in oil production, food processing, livestock feed	Warm, semi-arid, 25–35°C, sandy loam soils	Direct seeding, requires well-drained soil

Table 3.1 (Continued)

Name of crop (Scientific Name)	Country	Nutritional value	Economic value	Climatic suitability	Cultivation methods
Soybean (*Glycine max*)	Zambia, South Africa, Nigeria, Uganda	High in protein, omega-3 fatty acids	Used in animal feed, vegetable oil, and export markets	Tropical to temperate, 20–30°C, requires moderate rainfall	Direct seeding, nitrogen-fixing crop
Sesame (*Sesamum indicum*)	Ethiopia, Sudan, Uganda, Nigeria	Rich in healthy fats, protein, antioxidants	Used for oil extraction, high-value export crop	Hot, arid/semi-arid, 25–35°C, low rainfall	Direct seeding, drought-tolerant
Palm Oil (*Elaeis guineensis*)	Nigeria, DR Congo, Ivory Coast, Ghana	Contains vitamin E, antioxidants	Used in food, cosmetics, and biofuel industries	Humid tropics, 24–32°C, high rainfall	Grown from seeds, requires proper spacing
Sugarcane (*Saccharum officinarum*)	Egypt, South Africa, Mauritius, Sudan	Rich in sucrose, source of quick energy	Used in sugar production, ethanol, and beverages	Tropical, 25–35°C, high water requirement	Planted from cuttings, requires irrigation
Cotton (*Gossypium* spp.)	Mali, Burkina Faso, Benin, Chad	No direct nutritional value	High-value textile industry export crop	Warm, dry climates, 25–35°C, well-drained soils	Direct seeding, requires pest management
Banana (*Musa* spp.)	Uganda, Cameroon, Rwanda, Kenya	High in potassium, vitamin C, fiber	Exported fresh, used in food processing and flour	Humid tropics, 25–35°C, high rainfall	Grown from suckers, requires regular fertilization
Tomato (*Solanum lycopersicum*)	Egypt, Nigeria, Morocco, Sudan	High in vitamin C, lycopene, antioxidants	Used in sauces, canning, fresh market sales	Warm to temperate, 20–30°C, well-drained soils	Grown from seedlings, requires staking
Onion (*Allium cepa*)	Niger, Egypt, Ethiopia, Kenya, Burkina Faso	Contains vitamin C, antioxidants, anti-inflammatory	High demand in culinary industries	Semi-arid, 20–30°C, requires sandy loam soils	Grown from seeds/bulbs, requires proper spacing
Pineapple (*Ananas comosus*)	Ghana, Kenya, Ivory Coast, Benin	Rich in vitamin C, bromelain, aids digestion	Exported fresh, used in juice production	Tropical, 22–32°C, well-drained soils	Grown from crowns/suckers, requires mulching
Cashew Nuts (*Anacardium occidentale*)	Ivory Coast, Nigeria, Mozambique, Guinea-Bissau	High in healthy fats, protein, and minerals	High-value export crop, processed for oil and snacks	Warm, semi-arid, 25–35°C, requires deep soils	Grown from seeds or grafting, drought-resistant
Avocado (*Persea americana*)	Kenya, South Africa, Ethiopia, Tanzania	High in healthy fats, fiber, and vitamins	Exported for fresh consumption, used in cosmetics	Subtropical to tropical, 18–30°C, well-drained soils	Grown from grafted seedlings, requires pruning
Mango (*Mangifera indica*)	Mali, Kenya, Nigeria, Burkina Faso	High in vitamin A & C, antioxidants	High-value fresh fruit, processed into juices and dried fruit	Tropical, 24–35°C, well-drained soils	Grown from seeds/grafting, requires irrigation
Coconut (*Cocos nucifera*)	Tanzania, Ghana, Mozambique, Madagascar	High in electrolytes, healthy fats, and fiber	Used in oil production, cosmetics, and food industry	Coastal tropical, 25–35°C, saline-tolerant	Grown from nuts, requires sandy loam soil

Source: Compiled from FAO Crop Profiles, African Union Agricultural Reports, and country-level agricultural extension publications (2010–2023)

3.5 Livestock Farming

Limited writing exists on the contribution of livestock farming to the African landscape. It is worth noting that livestock farming is not so much an agricultural system but a predilection for keeping animals which does not depend on the absence of other opportunities for gaining better livelihood. This implies that much of livestock farming in Africa is less to do with an absence of crop land and resources, more so the desire to keep animals for cultural and social reasons. The keeping of animals dates back to the time when many of the pastoralist people moved into Africa. From as early as the Egyptians and Phoenicians, animals have been farmed from the Mediterranean across to the coast of West Africa. Since that time, there has been a vast diffusion of peoples and animals across the whole continent, resulting in a wide diversity of livestock systems and methods. The contribution of livestock farming to the African economy is significant as nearly 60% of Africans are involved in livestock raising activities. The recent growth in demand for animal products has increased the pressure on Africa to contribute to the rest of the world market, such as cattle (Fig. 3.6). This is even more paramount as the European Economic Community (EEC) makes ongoing adjustments to their Common Agricultural Policy (CAP) which directly influence trade agreements and tariffs affecting African country access to international markets.

3.5.1 Cattle

Cattle-herding practices in Africa are of great antiquity. Over the years, with the growth of mixed farming and evolving socio-economic dynamics,

Fig. 3.6 Herd of cattle (*Bos taurus*) in a ranch (Photo by E.T. Ababio)

the economic and social role of cattle has undergone significant transformations in various regions. Nevertheless, these magnificent creatures continue to play a vital and indispensable part in the farming systems across wide areas of the continent. This is especially true in the arid and seasonal regions of the north and south, where their unparalleled advantage as a store of realizable wealth during uncertain times and as a buffer in risk avoidance remains paramount. The types and breeds of cattle, along with traditional husbandry practices, have developed over time, adapting to changing circumstances and influenced by the rich tapestry of human and environmental history. Moreover, factors such as access to markets and the nature of contemporary farming systems have also left their imprints on these remarkable animals (Fig. 3.7). In marginal areas, where resources are limited, cattle can hold the key to efficient land use and conservation efforts (Fig. 3.8). Additionally, they serve as a valuable resource during peak labour shortages, offering support during crucial times of the year. However, it is essential to acknowledge that the presence of cattle can exert considerable pressure on other aspects of the farming system and environment (Fig. 3.9). Striking a delicate balance between the benefits they provide and the potential risks they pose is imperative. Therefore, comprehending the intricate correlation between cattle and farming systems is of utmost importance when considering any future changes or developments. By delving deeper into the multifaceted relationship between cattle and farming systems, we can gain valuable insights that will guide us towards sustainable practices and constructive growth. Such understanding will enable us to navigate the challenges that lie ahead, ensuring that the immense value of cattle in African agriculture is harnessed while safeguarding the well-being of the environment and the communities that rely on them.

Fig. 3.7 Herd of cattle grazing of a coastal savannah landscape (Photo by E. T. Ababio)

Fig. 3.8 Pictures of sheep (**A**) and goats (**B** and **C**) (Photo by E.T. Ababio)

Fig. 3.9 Cattle on grazing spree in a forest edge ecosystem (Photos by F. Annin)

3.5.2 Sheep and Goats

Sheep and goats (Fig. 3.8) were two of the earliest animals to be domesticated by humans in Africa. From around 6000 BC, an agricultural system developed whereby the cultivation of cereal crops (such as sorghum, pearl millet, and barley) was integrated with the keeping of livestock, the most common of which were small stock. This system was adopted throughout much of the sub-Saharan region since it was more suited to the arid conditions which started to spread at this time. Steeper declines in rainfall after about 1000 BC led to further growth of the system from the fringes of the Sahara into the savanna areas of West Africa. In recent centuries, small stock have generally been the most numerous type of livestock in the sub-Saharan region, albeit sheep and goats predominate in different ecological niches. Small stock can thrive in arid environments since they are browsers as well as grazers and so have a wider choice of foods than cattle or horses. Their relatively small body size means that they can utilize vegetation too low and sparse for cattle to consume, while their cloven hooves are less damaging to the environment than the hooves of horses or camels. Ruminant digestive systems allow efficient processing of high fiber, low-quality foods and the production of milk, another important product from small stock. Although the rationale behind the choice of sheep or goats in different places and periods is not always clear, historic and present distributions generally reflect ecological differences in the two animals' preferred habitats. Sheep are adapted to cooler, moister environments than goats and are generally subordinated to them in more arid areas.

3.5.3 Poultry

Poultry, also known as domesticated birds, have likewise played a crucial role in the development of human civilizations. The domestication of poultry species, such as chickens and ducks, can be traced back thousands of years. Similar to small stock, poultry animals have proven to be versatile and adaptable to various environmental conditions (Fig. 3.10). One of the main advantages of poultry farming is their ability to efficiently convert feed into protein-rich eggs or meat. This has made poultry an essential source of affordable and readily available animal protein in many regions around the world. Moreover, poultry animals require relatively small spaces compared to other livestock, making them suitable for both rural and urban settings. Their rapid growth rates and high reproductive capacities also contribute to their popularity in agricultural practices. In addition to their role as a source of food, poultry offer several other benefits. For instance, chicken feathers can be utilized in various industries, including clothing and bedding production. Poultry droppings, rich in nutrients, can be used as organic fertilizers, thus promoting sustainable agricultural practices. Furthermore, poultry animals exhibit behavioral traits that make them easy to handle and manage. They are generally social creatures, forming hierarchical structures within their groups. This social nature not only allows for efficient rearing and breeding practices but also contributes to their overall well-being. Over time, selective breeding and genetic advancements have led to the development of various specialized poultry breeds. These breeds are tailored for specific purposes, such as meat production, egg-laying, or ornamental qualities. As a

Fig. 3.10 A traditional poultry setting in rural Ghana (**A**) variety of birds (**B**) only local chicken (Photo by P.D. Agyemang)

result, the diversity of poultry species has grown tremendously, providing farmers and consumers with a wide range of choices. In conclusion, poultry animals have had a significant impact on human societies, contributing to food security, economic growth, and sustainable agricultural practices. Their adaptability, efficiency in feed conversion, and versatility make them an invaluable asset in the agricultural industry. As human populations continue to increase and global food demand rises, the importance of poultry farming is expected to grow even further.

3.5.4 Fish Farming

Fish farming is an agricultural practice that holds immense importance and has gained widespread popularity in West Africa due to the numerous benefits it brings to the region. The dominant form of fish farming in this area revolves around the cultivation and rearing of different species of tilapia in meticulously maintained ponds. In the country of Senegal, fish farming has assumed a critical role, utilizing traditional flood recession management techniques, which have proven to be tremendously successful in their extensive and low input practices, resulting in remarkable yields ranging between 200 and 500 kg/ha of fish. Although it may be considered less intensive compared to pond culture, which often involves the utilization of commercial fish pellets for feeding, this method still surpasses the productivity of natural fisheries in the region, where yields generally fluctuate between 20 and 200 kg/ha. One of the notable advantages of fish farming is its predictability and lower susceptibility to risks when compared to traditional agricultural practices. In regions where the pressure on land and its resources is consistently increasing, fish farming serves as a valuable source of both protein and income. Recognizing this potential, the commercial production of fish farming in West Africa frequently concentrates on urban areas where the demand for local protein sources is particularly pronounced. For example, in the nation of Ghana, the emergence of private fish farming in response to the decline of capture fisheries around the Volta reservoir has become an essential development. This decline is primarily attributed to overfishing and the escalating fishing efforts from migrant fishers. As a result, these private fish farming initiatives strive to meet the surging demand for fish in the local area as well as the bustling city of Accra (Fig. 3.11) (Table 3.2) (Abobi and Wolff 2020;

Fig. 3.11 A picture of fish pond on Lake Volta, Ghana (photo by E.A. Tetteh)

Table 3.2 Type of livestock farming, the countries involved, mode of feeding and socio-economic values in Africa

Farm type (Species Involved)	**Countries tnvolved**	**Farming system**	**Mode of feeding**	**Social value**	**Economic value**
Cattle Farming (*Bos taurus, Bos indicus*)	Kenya, Ethiopia, Nigeria, South Africa, Sudan	Nomadic pastoralism, Ranching, Mixed farming	Grazing (natural pastures), supplemented with hay, silage, or commercial feeds	Cultural status, dowry payments, religious sacrifices	Meat (beef), milk, hides & leather, manure for fertilizer
Goat Farming (*Capra aegagrus hircus*)	Nigeria, Mali, Ghana, Ethiopia, South Africa	Extensive pastoralism, Agro-pastoralism, Backyard farming	Browsing (shrubs, leaves), supplemented with grain	Household savings, dowry payments, traditional feasts	Meat (chevon), milk, leather, manure
Sheep Farming (*Ovis aries*)	Sudan, Somalia, Ethiopia, Morocco, Algeria	Pastoralism, Semi-intensive, Integrated farming	Grazing on pastures, supplemented with grains	Used in religious sacrifices, community feasts	Meat (mutton), wool production, milk
Poultry Farming (*Gallus gallus domesticus, Numida meleagris* - Guinea fowl)	Ghana, South Africa, Nigeria, Egypt, Kenya	Free-range, Semi-intensive, Intensive commercial	Commercial feeds, kitchen waste, grains	Used in festive celebrations, social gatherings	Eggs, meat, feathers, manure
Camel Farming (*Camelus dromedarius, C. bactrianus*)	Somalia, Sudan, Ethiopia, Mauritania, Chad	Nomadic pastoralism, Semi-intensive	Browsing on desert vegetation, salt licks	Symbol of wealth, transportation, cultural heritage	Meat, milk, leather, tourism (camel riding)
Pig Farming (*Sus scrofa domesticus*)	Uganda, Nigeria, South Africa, Cameroon, Kenya	Intensive piggery, Free-range farming	Commercial feeds, swill feeding, pasture-based	Used in traditional feasts, dowry payments	Meat (pork), lard, hides
Rabbit Farming (*Oryctolagus cuniculus*)	Kenya, Ghana, Nigeria, South Africa, Uganda	Smallholder backyard farming, Semi-intensive	Fresh vegetables, pellets, hay	Used in small-scale backyard farming, children's pets	Meat, fur, manure
Beekeeping (*Apis mellifera* - Honeybees)	Ethiopia, Tanzania, Kenya, South Africa, Nigeria	Traditional beekeeping, Modern apiary farming	Forages on flowers for nectar, artificial feeding in dry seasons	Used in traditional medicine, spiritual significance	Honey, beeswax, pollination services
Fish Farming (*Tilapia - Oreochromis* spp., Catfish - *Clarias* spp.)	Egypt, Ghana, Nigeria, Uganda, Zambia	Aquaculture (ponds, cages, tanks)	Formulated pellets, natural pond feeding	Source of food in local diets, traditional fishing practices	Freshwater fish production, export, aquaculture business

Table 3.3 (Continued)

Farm type (Species Involved)	**Countries tnvolved**	**Farming system**	**Mode of feeding**	**Social value**	**Economic value**
Donkey Farming (*Equus africanus asinus*)	Ethiopia, Burkina Faso, Niger, Chad, Sudan	Pastoralism, Mixed farming	Grazing on open fields, occasional grain supplements	Used for transportation, plowing fields	Labor (draft power), hides for export (Ejiao production)
Horse Farming (*Equus ferus caballus*)	Morocco, Egypt, South Africa, Nigeria, Sudan	Free-range grazing, Stall-fed systems	Grazing, supplemented with grains, hay	Used in sports, cultural parades, status symbol	Transportation, racing, tourism
Duck Farming (*Anas platyrhynchos domesticus*)	Nigeria, South Africa, Egypt, Uganda, Cameroon	Free-range, Semi-intensive	Free-range foraging, supplemented with grains	Used in festivals, family nutrition	Meat, eggs, feathers

Araujo et al. 2022; Kamagaté et al. 2021; Ainsworth et al. 2021; Musa 2023; Barthélemy et al. 2024; Munguti et al. 2022; Beg et al. 2024; Sinhababu and Poonam 2022; Chang et al. 2020).

3.6 Challenges in African Agriculture

With the reverse engineering of the last 50 years, starting from independence, we can deduce many challenges for the future development of African agriculture. Today, facing economic liberalization, globalization, and the global knowledge-based economy (GoKE), the agricultural economy has sustained very low growth, making the sector problematic. Low growth implies low agricultural research and development (R&D); the most important source for medically, nutritionally, and economically improving society and the poor.

The low growth in the African agricultural sector can be attributed to several different factors. Firstly, government investment in agricultural R&D is low. Over the last 20 years, the CGIAR organization and public sector institutions have seen investment slump from 3.1% to 1.8%, yet improved the quantity and quality of life for farm households by an estimated fifty percent. It is clear that the government should view agricultural R&D as a solution, not a cost, and increase investment.

The second and most important factor hindering growth is poor policy making and governance. This can manifest as high taxes on export crops, favoritism to urban areas, anti-agricultural bias, bureaucracies, and corruption. Theoretically, improved governance would improve the quality and relevance of public sector priorities and increase decision-making efficiency, thereby improving the quality of lives in Africa. Unfortunately, corruption is a hard trait to eliminate due to its transferability between generations and its seen as a shortcut method to high social rankings and monetary gains. Corrupt behavior can be seen as a rational response to the low salaries in public sector administration, yet it will continue to hinder R&D production and social welfare (Pingali 2007).

A third factor is the removal of trade barriers, which may expose African farmers to serious competition from other countries, and the marginalization of environmental management by the poor in the context of overall economic growth could result in the loss of resources for the benefit of future generations. All the aforementioned problems decipher one thing: current policies are growing a large disconnection between small farmers and agricultural R&D and new technology.

3.6.1 Climate Change

Recent or projected climatic changes at the global level are having, and could continue to have, an important and profound impact on African agriculture. This is primarily due to the fact that the vast majority of farming practices in Africa heavily rely on rainfall, as only a mere 4% of the continent's cropped land is equipped with irrigation systems. It is crucial to note that drought-prone semi-arid areas, which are home to approximately 64% of the continent's population, are particularly vulnerable to these unpredictable weather patterns. Given the current trajectory, it is widely anticipated that climate change will bring about substantial increases in temperature across many regions, accompanied by significant alterations in precipitation levels and the frequency of extreme weather events. Nevertheless, while the former is widely acknowledged, there remains a certain degree of uncertainty surrounding the latter. As a consequence of these changes, pastoralism and extensive cropping, particularly in regions such as East Africa and the Sahel, are expected to face immense challenges and hardships under numerous future climate scenarios. To gain a deeper understanding of the predicament at hand, it is vital to examine concrete and tangible examples of the issues faced by African agriculture. A seminal piece of research provides an exhaustive and comprehensive examination of the vulnerability of East African arid and semi-arid lands (ASALs) in relation to climate change. The study meticulously analyzes the intricate details of the challenges encountered by communities in these regions, shedding light on the severity and complexity of the situation. Undoubtedly, the implications of climate change on African agriculture cannot be underestimated. There is an urgent need for proactive measures to be taken at both local and international levels in order to mitigate the adverse effects and enhance the resilience of vulnerable farming communities. Collaboration between governments, researchers, and local stakeholders is paramount to implementing effective adaptation strategies and ensuring the long-term sustainability of African agriculture in the face of an increasingly uncertain climate.

3.6.2 Limited Access to Resources

Because of the critical and significant nexus between soil fertility and population density, it is reasonable to anticipate that African farmers would exhibit a high degree of conscientiousness when it comes to the preservation and upkeep of their land. This assumption holds true for many regions across Africa; however, in the expansive and diverse ecological landscapes of savannas and forests, the dynamic of frequent migration has emerged as the predominant response to fluctuating levels of soil fertility, tsetse fly risks, and political circumstances. In this context, the principal obstacle to making long-term investments in land quality boils down to one central factor: insecurity. Whenever land ownership becomes volatile and precarious due to the looming specter of war, ever-shifting political alliances, or drastic changes in national policies regarding land tenure, farmers naturally exhibit very little enthusiasm or inclination to enhance the land they possess. At most, they might resort to clearing forested areas, burning the vegetation in order to release valuable nutrients into the soil, and plant crops for a limited number of seasons. The subsequent natural and gradual increase in soil organic matter helps improve fertility to some extent. Yet, as soon as the soil fertility begins to dwindle, these farmers swiftly make the decision to shift to a different location. This distinct pattern of land utilization often carries the label of "slash-and-burn" agriculture, although it proves to be more reflective of the severely restricted and tightly bound modes of land use than a purposeful and intentional strategy aimed at preserving land in a fallow state over prolonged periods of time. The implications of this pattern of land use are far-reaching. Not only does it hinder the long-term sustainability of agricultural practices, but it also perpetuates a cycle of soil degradation, low productivity, and reduced food security. Furthermore, the migration of farmers in search of fertile land contributes to population displacement, increased competition for limited resources, and social upheaval. As such, finding viable solutions to encourage African farmers to adopt more sustainable approaches to land use and management is of utmost importance. One potential avenue for addressing this challenge is through the provision of secure land tenure rights. Ensuring that farmers have legal ownership or access rights to the land they cultivate can provide them with the necessary incentives to invest in its long-term improvement. By having a sense of permanence and stability in their land tenure, farmers may be more inclined to implement sustainable practices such as agroforestry, terracing, rotational grazing, and the use of organic fertilizers. These practices can help enhance soil fertility, reduce erosion, promote biodiversity, and ultimately bolster agricultural productivity. However, securing land tenure rights alone may not be sufficient to bring about the desired changes in land use patterns. It is crucial to accompany this with comprehensive support and education programs that empower farmers with the knowledge and skills needed to adopt sustainable land management practices. Training programs can focus on soil conservation techniques, water management, integrated pest management, and climate-smart agriculture. By equipping farmers with the necessary tools and information, they can make informed decisions about their land use practices and embrace more environmentally friendly and economically viable approaches. Additionally, fostering collaboration and knowledge-sharing among farmers, researchers, policymakers, and agricultural experts can play a significant role in promoting sustainable land use in Africa. Through platforms such as farmer field schools, agricultural extension services, and community-based organizations, valuable insights and best practices can be shared, enabling farmers to learn from each other's experiences and adapt innovative techniques to their local contexts. This collaborative approach can also facilitate the adoption of inclusive and participatory land governance models that ensure the voices and interests of small-scale farmers, indigenous communities, and marginalized groups are effectively represented in decision-making processes. Ultimately, achieving sustainable land use in Africa requires a multi-faceted and holistic approach that addresses the complex interplay of social, economic, and environmental factors. By combining secure land tenure rights, targeted education and training programs, and collaborative platforms for knowledge exchange, African farmers can be empowered to break free from the shackles of insecure land ownership and embrace practices that not only enhance soil fertility and agricultural productivity but also contribute to broader goals of social stability, food security, and environmental sustainability.

3.6.3 Pest and Disease Management

Farmers in Africa have a wide range of diverse and multifaceted options available to them when it comes to mitigating and managing pests and diseases affecting their crops. These options encompass both modern chemical and physical technologies, as well as traditional and indigenous methods of pest and disease control. However, it is crucial to acknowledge that a significant proportion of impoverished farmers face constraints in their ability to effectively combat numerous pests and diseases, necessitating further improvements in their management strategies. This need for enhanced methods of control is particularly pertinent in the realm of crop improvement programs. These programs consistently emphasize the development and cultivation of resistant cultivars capable of withstanding the damaging impact of specific pests or diseases. This approach is widely regarded as economically prudent and sustainable in the long run, proving to be a cost-effective means of maintaining desired levels of control. Nevertheless, it is crucial to contextualize disease and pest control within the broader framework of the entire farming system. In regions impeded by severe production constraints, the immediate objective revolves around rehabilitating the farming system to transform it into a viable and substantial food producer. Consequently, the methods employed for pest and disease control in these situations must be capable of facilitating the re-establishment of food production without relying on costly inputs. Such an approach allows for the eventual growth of farm profitability, which paves the way for the exploration of alternative control methods that may not have been previously viable or feasible. Throughout the pest and disease control process, it is imperative to avoid inadvertently creating an ecological niche that could foster the development and proliferation of other pests or diseases. An illustrative example involves the suppression of one pest species that acts as a natural predator to a second, more harmful pest. This suppression can inadvertently create a vacuum that encourages the emergence and intensification of additional pest problems. Therefore, careful consideration must be given to selecting the most suitable method of control, taking into account the potential ecological repercussions and the long-term consequences associated with each choice. By adopting a comprehensive and strategic approach to managing pests and diseases, African farmers can safeguard their crops, maximize productivity, and ensure sustainable food production. The continuous exploration and implementation of innovative and environmentally conscious solutions are imperative for the long-term success of agricultural systems across the continent.

3.6.4 Market Access

Thus far we have meticulously examined and delved into the intricate and intricate complexities of the smallholder agriculture sector in Southern Africa. We have expounded upon the extensive, multifaceted nature of impoverishment, exhaustively exploring its extent, causes, and underlying factors. In light of these contemplative discussions, we now turn our attention to the pivotal questions that shall shape the remainder of this paper: Does participation in smallholder farming perpetually propagate poverty as the sole outcome? And, perhaps more importantly, are there sustainable and feasible farming alternatives that can serve as a means of emancipation from the clutches of destitution? The significance of these inquiries cannot be understated, as the issue at hand is both profound and multifaceted. Pursuing a singular focus on the promotion of commercial agriculture in the region does not necessarily guarantee a reduction in poverty, primarily due to the prevalent prevalence of casual labor within this geographical domain. It is critical that we carefully analyze the intricate dynamics at play in order to ascertain the potential pathways that lead individuals and communities towards a life free from the shackles of poverty. Therefore, we endeavor to conclude that there exist numerous viable trajectories that can serve as an escape from the perils of smallholder farming-induced impoverishment. Our examinations will lead us to advocate for the implementation of successful strategies centered around crop diversification and livelihood enhancement, which we deem to be pivotal in achieving sustainable poverty alleviation. It is essential to acknowledge that the trajectory of agricultural production in this region has been significantly shaped by the market-oriented policies adopted by African countries over the past three decades, both in historical retrospective and the present day context. Throughout this period, the implementation and widespread adoption of the green revolution technologies have remained notably scarce and neglected south of the vast Sahara. Concurrently, the global landscape witnessed the advent of market liberalization, which resulted in the majority of countries retracting government subsidies previously extended to smallholder farmers, while simultaneously enforcing and perpetuating low producer prices. The cumulative effects of these comprehensive policy frameworks created a disheartening environment that provided limited incentives for smallholders to augment their production capacities. Consequently, the vulnerability of households to impoverishment escalated drastically, with dire consequences affecting the broader societal fabric. Furthermore, this grim scenario was further aggravated by the breakdown of formal market systems that bedeviled numerous countries within the region. The overvaluation of currency and the zealous enforcement of Structural Adjustment policies acted as catalysts in this destructive process, leading to the dismantlement of agricultural marketing boards and the disintegration of the formal sector. The forces of liberalization ushered in an inundation of inexpensive imports, which, in turn, prompted a notable shift in the orientation of smallholder production towards lower-value subsistence crops. As a lamentable consequence, the volume of products sold in the market drastically declined, eliciting widespread market segmentation and a palpable rupture between liberalized and globalized markets for cash crops and strategic food staples. Unsurprisingly, this yielded a regrettably ubiquitous state of food insecurity and a pervading sense of poverty that loomed over communities like an ominous specter. The repercussions of these adverse circumstances have been both dire and far-reaching, culminating in the manifestation of extreme market fragmentation. This multifaceted phenomenon has decisively contributed to the perpetuation of poverty on an alarming scale and created an expansive divide between those who have easy access to liberalized markets and those who are confined by the constraints of strategic subsistence-oriented agriculture. The consequences of this fragmentary state of affairs are severe and all-encompassing, engendering a distressing state of food insecurity and debilitating poverty that afflict the region at large (Clay and Zimmerer 2020; Rohne Till 2021; Boafo and Lyons 2022; Abdulai 2022; Jellason et al. 2021; Dahunsi et al. 2020; Ochieng et al. 2022; Francisco Ribeiro and Camargo Rodriguez 2020; Hurt 2020; Harwood 2020).

3.7 Agricultural Policies and Initiatives

3.7.1 Government Support for Farmers

Various African governments have provided extensive support and resources to farmers in order to significantly enhance and broaden their agricultural production. This immense level of support has been influenced by a wide range of evolving agendas, partly influenced by prevailing development theories and trends, and often driven by political leaders' endeavors to gain favor with influential rural constituents. The support has materialized primarily through the facilitation of rural credit provision and the implementation of favorable price policies. For instance, in Nigeria, until the mid-1980s and even during the most recent structural adjustment program, the government operated commodity marketing boards that took the responsibility of purchasing crops from farmers at fixed prices. These fixed prices had substantial appeal for farmers, who were aware of the significant profit margins they could obtain by investing in fertilizers and acquiring tractor equipment. However, in most cases, the surplus generated from these agricultural endeavors ended up being consumed by middle-class traders who hugely benefited from the heavily protected domestic commodity markets. As a result, governmental policies have also been designed to support the "rural bourgeoisie" and their efforts to intensify agricultural practices. This approach has been perceived as an important step towards establishing capitalist relationships within the agricultural sector. Conversely, small-scale farmers with limited capital have often been marginalized and compelled to rely on rural subsistence activities. A typical example of this occurred in Kenya during the early 1980s, when the ruling elite, under the leadership of Jomo Kenyatta, exploited the opportunities presented by "preferential" exchange rate policies and lucrative producer prices for export crops. This allowed them to accumulate substantial capital through extensive coffee production in fertile regions. Unfortunately, this process of modernization came at the expense of African smallholder producers losing access to their land and experiencing additional marginalization through the recruitment of landless individuals for plantation wage labor. In a parallel development during the same era, landlords in the densely populated Central African region of Malawi established connections with local government officials. This enabled them to circumvent the procedures for public land acquisition and acquire vast tracts of land in customary areas that were originally designated for smallholder agriculture. In some cases, "communal" farmers were not even given prior notice of this land loss and only became aware of the land alienation when immigrant labor tenant families, who had been promised both wage work and land tenancy by elite farmers, violently demanded resources on the smallholders' farmland. This movement of resources and capital to marginalize the peasantry in Africa serves as a critical theme in the ongoing modernization processes that are profoundly shaping the African agricultural landscape. This trend highlights the complex dynamics between political leaders, rural constituents, and the agricultural sector, showcasing the challenges faced by small-scale farmers and the potential discrepancies in the distribution of resources and power. It is essential for African governments to strike a balance between supporting the growth of agricultural production and ensuring equitable opportunities for all farmers, regardless of their economic status or land ownership. Only through comprehensive and inclusive policies can the agricultural sector in Africa truly flourish, leading to sustainable development and improved livelihoods for rural communities (Jayne and Sanchez 2021;Onyiriuba et al. 2020; Santpoort 2020; Mapiye et al. 2021;

Langyintuo 2020; Sekaran et al. 2021; Giller et al. 2021; Bjornlund et al. 2020; Giller 2020).

3.7.2 Investment in Agricultural Research and Development

A more recent and noteworthy development in augmenting the agricultural potential of the vast continent of Africa has been the growing investment in agricultural research and development. It is widely acknowledged that public funding on agricultural research and development (R&D) has, for the most part, fallen short in producing substantial results that can make a significant impact on agricultural productivity. As a result, various international agencies have stepped in, fueled by the intention to increase aid fourfold within the next decade, with an annual expenditure reaching a staggering 1 billion. Among these influential international agencies, the consultative group on international agricultural research (CGIAR) stands out. Established in 1971, CGIAR's primary goal was to eradicate world poverty and malnutrition through strategic and high-quality agricultural research. Over the past 30 years, CGIAR has made commendable strides in improving the livelihoods of people in developing countries through groundbreaking agricultural research. While Africa has received a substantial portion of this aid, there is a valid argument that the research conducted by CGIAR needs to be more efficient and focused in addressing African agricultural issues. Consequently, negotiations regarding CGIAR's reform have recently taken place, aiming to ensure a more stable funding structure and a greater allocation of resources directly towards African agricultural research. However, critics contend that there is an overreliance on input-intensive research and an excessive influence from Western donor countries within CGIAR. These critics advocate for a more in-depth empirical exploration of African farming systems and a comprehensive understanding of the constraints faced by farmers on the continent. Furthermore, it is strongly argued that the privatization of agricultural research is necessary to align research with market demands and ensure efficient resource management. This proposal is grounded in the shortcomings of publicly funded research, which has sometimes failed to deliver satisfactory results. Nevertheless, it is important to acknowledge that not all public research is unproductive, and it cannot be assumed that private organizations will readily invest the necessary funds and effort to address the specific challenges faced by smallholder farmers. Interestingly, there has been an emerging trend of large seed and chemical companies taking on the considerable risk and investing in R&D in Africa. This trend is primarily driven by the fact that 80% of the world's farmers are classified as smallholders, and it is projected that within 40 years, the majority of these farmers will become consumers of generic farm inputs. While investment in R&D undeniably holds the potential to reverse the decline in African productivity, it must be recognized that it is an inherently risky endeavor. Moreover, there is a legitimate concern that such investment may lead to increased inequality among African farmers. In conclusion, the increasing investment in agricultural research and development in Africa represents a promising step towards unlocking the continent's agricultural potential. However, there is a pressing need for greater efficiency, focus, and reform in agricultural research efforts. By addressing the unique challenges of African farming systems, promoting demand-driven research, and adopting prudent resource management strategies, the goal of a "Green revolution" in Africa can be realized. Nonetheless, caution must be exercised in navigating the delicate balance between risk-taking investment and the potential implications it may have on the equitable distribution of resources among African farmers. The increasing investment in agricultural research and development in Africa represents a promising step towards unlocking the continent's agricultural potential. However, there is a pressing need for greater efficiency, focus, and reform in agricultural research efforts. By addressing the unique challenges of African farming systems, promoting demand-driven research, and adopting prudent resource management strategies, the goal of a "Green revolution" in Africa can be realized. Nonetheless, caution must be exercised in navigating the delicate balance between risk-taking investment and the potential implications it may have on the equitable distribution of resources among African farmers (Byerlee and Lynam 2020; Byerlee and Lynam 2020; Plucknett 2020; Stapleton 2022; Lynam et al. 2024).

3.7.3 International Aid and Partnerships

International aid, including donor funding and partnership projects, has played a major role in shaping agriculture and addressing food security across the continent. Following independence, many African governments chose to focus on food crop production, often at subsistence level, in order to achieve food self-sufficiency. This was generally not successful due to the poor post-harvest infrastructure and limited commercial opportunities in many remote rural areas, combined with a decline in the price of many food crops compared with export commodities. Since the early 1980s, structural adjustment programmes imposed by the International Monetary Fund (IMF) and World Bank on heavily indebted countries led to a greater emphasis on liberalisation of agricultural policy, as well as reduction in government spending in the agricultural sector. This has often increased social and economic differentiation within rural areas, as well as levels of relative poverty. In recent years, the continent has seen a major increase in food aid in response to widespread famine brought about by drought and poor governance, particularly in southern and eastern Africa. Partnerships, particularly those involving one-to-one country relationships with those in Western Europe, have been a major focus of aid for agricultural development, seeking to harness private sector investment and trade liberalisation agreements. By far most substantial of these has been the relationship between the ACP group of African, Caribbean and Pacific countries and the European Union (EU), which dates back to the Rome Treaty of 1957. This has included provisions for preferential trading agreements, structural funding, and budgetary aid, as well as the more recent Cotonou Agreement which aims to eradicate poverty and achieve sustainable development. On a smaller scale, there has been a considerable amount of bilateral partnership work between African states and individual European countries, as well as the United States of America (USA). One of the major current issues in development partnerships is the involvement of emerging donor nations such as China, India, and Brazil, seeking to establish a coalition of the South in contrast to the traditional aid model imposed by the North, and to further their own agricultural interests in the Global South. These emerging donor nations have recognized the potential impact of their investments and have positioned themselves as key players in the agricultural development landscape, bringing new perspectives and resources to the table (Watts and Scales 2020; Smidt and Jokonya 2022; German et al. 2020; Antwi-Agyei et al. 2021; Jayne et al. 2021).

3.8 Role of Women in Agriculture

In many parts of Africa, farming is largely, if not entirely, a woman's domain. Women are actively involved in various tasks related to agriculture, such as weeding, planting, and harvesting. Moreover, female farmers often work independently as well as in groups, and there is strong evidence to suggest that women's groups can be more efficient than male groups in certain agricultural activities. The contributions of women toward agricultural production in Africa are undeniably considerable. In fact, in some areas of Africa, women put in significantly more hours into farming than men. This is particularly true in regions with high male out-migration. For instance, East African women have been reported to dedicate 60 to 80 percent of their labor to producing food crops. Similarly, in parts of Lesotho, women claim to carry out 50 to 80 percent of the plowing, even when men are present. The dedication and involvement of women in farming are undeniable. However, it is essential to address the prevalent interpretation of women's work as primarily or solely supplemental to that of men. This misleading interpretation stems from the common practice of ignoring or undervaluing women's substantial contributions to agricultural activities. Agricultural activities hold immense social and symbolic significance for African women. Women have a deep association with wet cultivation and the earth, which are intricately linked to fertility. As a result, women are often viewed as the symbolic and actual guardians of the earth and its reproductive and regenerative functions. Notably, women's labor in the farming sector is frequently intertwined with the crucial need to feed the family. Consequently, the protection of women's rights to land and resources for farming and their families becomes inherently linked to their ability to provide sustenance. In a patriarchal society, a woman who is unable to fulfill her duty of feeding her family due to a lack of access to resources may face the denial of her rights to those very resources. This further exacerbates her vulnerable position within the household. An enlightening study conducted in Uganda revealed that men were increasingly abandoning agricultural activities in the pursuit of paid employment. Consequently, women found it easier to assert their own authority over farming, especially on land that had been abandoned by men. However, it is crucial to emphasize that the ability of women to exercise control over specific resources for farming and ensure long-term security is contingent upon various local circumstances. Typically, women gain access to resources by overpowering male control, forming alliances with supportive men, or receiving specific resource rights through inheritance or land titles. These circumstances play a vital role in determining the extent to which women can exercise control and secure their rights over essential resources for successful farming. Overall, women's indispensable role in agriculture is not just limited to their immense labor and extended working hours. It extends to their vital contributions to the social fabric, fertility, and sustainable livelihoods within their communities (Pierotti et al. 2022; Agarwal 2022; Oppong 2023; Glazebrook et al. 2020; Fonjong and Gyapong 2021; Schildkrout 2023; Abu 2023; Kirkwood 2022; Ball 2020; Danso-Abbeam et al. 2020).

3.8.1 Women as Farmers

Sub-Saharan Africa remains one of the foremost preliterate areas of the world. Here, the agricultural scene is characterized by a rich diversity of crops ranging from indigenous rain-fed cereals such as millets, sorghum, and teff through wheat, maize, and rice, to highly commercialized export crops like coffee, cocoa, and cotton. It is also very much an Africa of small-scale farmers growing crops on landholdings of between 1 and 10 hectares. Set in the context of population growth rates that are still among the highest in the world, the African agricultural arena is of pivotal importance to the continent's most pressing contemporary problem: how to ensure food security for a growing population. In this thought-provoking essay, a comprehensive and broad characterization of the rural agricultural landscape is provided. The first part meticulously sets the scene by thoroughly examining the intricate biophysical and historical context. This extensive exploration creates a robust framework for better understanding the immensely diverse agricultural systems described in the second part. The final and most insightful section delves into the complex issues of macro-economic policy, laser-focused on agriculture in the African context of structural adjustment and associated liberalization. The central argument put forth in this profound essay is that for a complete and comprehensive understanding of agricultural diversity in Africa, it is of utmost importance to refocus attention upon the often overlooked small-scale farming sector. Unfortunately, the prevailing preoccupation with the constraints of the export sector has led to an oversimplified categorization of African agriculture as "dualistic". However, this oversimplification fails to capture the astonishing complexity and vast array of farming systems that exist in Africa. In reality, what may come across as 'traditional' agriculture is often a deeply rational adaptation to severe agroecological constraints. Moreover, dualistic models categorically fall short in explaining the intricate web of factors that have shaped and influenced African agriculture into its current state. As we delve further into the remarkable world of African agriculture, we begin to appreciate the interplay of numerous factors, both historical and contemporary, that have shaped this dynamic and fascinating landscape. By embracing a holistic perspective that considers the inherent complexities, this essay aims to shed light on the true essence of African agriculture. It unveils the intricate tapestry of small-scale farming systems that have taken centuries to evolve, balancing tradition with innovation, and showcasing remarkable resilience in the face of immense challenges. In conclusion, Sub-Saharan Africa stands as a testament to the extraordinary diversity and vibrancy of agricultural practices. By recognizing the crucial role of small-scale farming and acknowledging the multifaceted nature of African agriculture, we take a significant step towards addressing the urgent issue of food security for a rapidly growing population. A profound understanding of the complexities at play in African agriculture empowers us to formulate inclusive and robust policies that can propel this vital sector towards sustainable development and prosperity, benefiting not only the continent but the world at large. The future of African agriculture lies not only in the hands of policymakers and economists, but also in the grassroots efforts of farmers themselves, who hold a wealth of traditional knowledge and innovative solutions. Together, we can build a brighter, more resilient, and equitable agricultural future for Africa, ensuring that its rich diversity continues to flourish for generations to come.

3.8.2 Women in Agricultural Leadership

Women in leadership or management positions in African agriculture have received relatively limited attention from researchers, government officials, development agencies, and farmers. This contrasts with the overwhelming evidence that women are major contributors to agricultural production in Africa. A 1990 study of 24 African countries showed that women contribute 60–70% of labor in food production, while in some regions in Eastern and Southern Africa, women provide up to 90% of labor for certain crops. Despite the quantity and quality of their work, though, women continue to be marginalized from the mainstream of agricultural development. Their low socioeconomic status, heavy workload, and limited access to resources have prevented women from taking on more leading roles in agriculture. Nevertheless, there are many women in Africa who have managed to overcome these obstacles and assumed control of major responsibilities or leadership roles in agriculture—as farmers, community leaders, entrepreneurs, innovators, or civil servants. An exploration of these women and how they managed to achieve their positions can provide valuable insight. Leadership is difficult to define and even more difficult to measure. What makes a good leader? How can leadership be quantified? Should we measure leadership by the extent of a person's control over resources? By the ability to influence and motivate others? By the effect that a person's actions have on the overall direction of things? All of the above are aspects of leadership. When we refer to women in leadership in African agriculture, the most common assumption is that leadership is depicted by control over resources. This is especially relevant in a continent where land is a key asset and indicator of status. For this reason, many studies of women in leadership in African agriculture have focused on women who have secure rights to land. Yet women who are actively involved in farming and manage to support a family off the produce of their own or spouses' plots, which can be common in subsistence agriculture in Africa, are in effect leaders by the earlier definition in that they are contributing greatly to the direction of agricultural production and the well-being of the family. These women often face additional challenges such as limited access to education, markets, and technology, which further hinder their ability to fully unleash their leadership potential. Therefore, it is crucial to recognize the diverse forms of women's leadership in African agriculture and to provide them with the necessary support and resources to thrive. Additionally, it is important to acknowledge that women's leadership in African agriculture extends beyond their roles as farmers. Women are also making significant contributions as community leaders, entrepreneurs, innovators, and civil servants. They are actively involved in decision-making processes, advocating for their rights, and implementing sustainable agricultural practices. By actively participating in various sectors of the agricultural value chain, women are driving positive change and transforming rural economies. Their leadership is not limited to the confines of their own farms but extends to influencing policies, strategies, and development agendas at local, national, and international levels. Recognizing and promoting the leadership of these women is not only essential for gender equality and social justice but also for addressing the complex challenges facing African agriculture and achieving sustainable food systems. In conclusion, women's leadership in African agriculture is multi-dimensional and encompasses various roles and responsibilities. Despite the numerous barriers and societal norms that hinder their progress, women are actively contributing to agricultural production, community development, and sustainable resource management. Their leadership is crucial for achieving food security, poverty reduction, and overall socio-economic progress in Africa. Therefore, it is imperative to invest in programs and initiatives that empower and support women in agriculture, providing them with access to education, finance, technology, and markets. By doing so, we can unlock the full potential of women's leadership and foster inclusive and sustainable agricultural systems that benefit all.

3.9 Technological Advancements in Agriculture

agriculture, making it more resilient and productive. In recent years, the importance of sustainable farming practices has gained traction in Africa. Efforts have been made to promote agroecology, which emphasizes the use of natural processes and biodiversity to maintain soil fertility and combat pests. Through agroecological methods, farmers are able to achieve higher yields while minimizing the use of external inputs such as synthetic fertilizers and pesticides. This shift towards ecological farming has not only improved the sustainability of African agriculture but has also enhanced food security and livelihoods. Furthermore, advancements in information and communication technology (ICT) have revolutionized the agricultural sector in Africa. Mobile phones and internet access have enabled farmers to access vital information on weather patterns, market prices, and best farming practices. This has empowered farmers to make informed decisions and adapt to changing conditions, resulting in increased productivity and profitability. Additionally, e-commerce platforms have emerged, connecting farmers directly to consumers and eliminating intermediaries, thus increasing their income and reducing post-harvest losses. Another significant development in African agriculture is the adoption of climate-smart practices. As climate change poses a growing threat to agricultural productivity, farmers have been encouraged to employ techniques that mitigate its impact. These include conservation agriculture, which involves minimal soil disturbance, diversified crop rotations, and the use of organic inputs. By implementing climate-smart farming methods, African farmers are better equipped to cope with extreme weather events and ensure the long-term sustainability of their livelihoods. In recent years, there has also been a surge in the use of precision agriculture techniques in Africa. Through the application of remote sensing technologies, such as satellite imagery and drones, farmers are able to monitor crop growth, detect diseases, and optimize the use of inputs on a precise and localized scale. This not only improves resource efficiency but also reduces the environmental footprint of agriculture. Furthermore, the use of data analytics and machine learning algorithms enables farmers to make data-driven decisions and enhance their overall productivity. In conclusion, African agriculture has undergone remarkable transformation since the 1960s, driven by advancements in technology and a shift towards sustainable practices. These developments have increased the resilience and productivity of the sector, while improving food security and livelihoods for millions of farmers. With ongoing innovations and investments, the potential for further growth and development in African agriculture is immense, paving the way for a prosperous and sustainable future agriculture.

3.9.1 Sustainable Solutions for Food Security

One of the most critical issues regarding Africa's food security is sustainability, both in rural and urban areas. Some 200 million people rely exclusively on shifting cultivation to meet their food needs, with another 200 million practicing it to some extent. The population densities on these lands are usually very low - one to four persons per square kilometer—and the soil is quickly exhausted because the natural vegetation is burned and because the length of the fallow period has decreased. In extreme cases, the land is abandoned after one year in favor of clearing another plot and the original plot is left to revert to bush. A more sustainable way of using the land is needed to feed Africa's rapidly growing population. Unfortunately, economic and demographic pressures often result in short term economic gains being given priority over long-term sustainability. This is also manifested in land tenure policies which encourage rapid depletion of resources in order to prevent other social groups from encroaching on a particular area of land. High population growth rates make it difficult for "land-rich, people-poor" communities to maintain extensive fallow periods and also force people to marginal lands and fragile environments. The Green Revolution had varying effects in Africa, but in general it has led to greater differentiation between those who have access to productive lands and those who do not and has decreased the amount of staple food that is grown in many areas. This increasing integration between local, national, and international markets has both positive and negative effects but may not be desirable for some isolated rural communities. It is crucial to address these challenges and prioritize long-term sustainability in Africa's food security efforts. Sustainable farming practices that focus on replenishing the soil, preserving natural vegetation, and promoting extensive fallow periods should be adopted. Additionally, land tenure policies should be revised to prioritize sustainable resource management and ensure equitable access to productive lands. Efforts should also be made to support "land-rich, people-poor" communities, providing them with the necessary resources and knowledge to maintain sustainable agriculture practices. This could include educational programs, access to modern farming techniques, and financial support for investments in agricultural infrastructure. Furthermore, while the Green Revolution has brought about positive change, it is important to critically evaluate its impact and adapt strategies accordingly. The focus should be on empowering local communities, promoting self-sufficiency, and ensuring that the benefits of agricultural development reach all segments of society. Overall, achieving food security in Africa requires a comprehensive approach that prioritizes long-term sustainability, equitable resource management, and the empowerment of local communities. By addressing these challenges head-on, Africa can work towards a future where its growing population is well-nourished, its lands are productive and resilient, and its food systems are sustainable for generations to come.

3.9.2 Promoting Youth Participation in Agriculture

A comprehensive, in-depth review was conducted, delving deep into the rich experiences and endeavors of various African countries in the tireless pursuit of fostering and enhancing youth participation in the dynamic realm of agriculture. The countries under scrutiny were none other than the remarkable nations of Ghana, Mali, Rwanda, Sao Tome & Principe, Senegal, and Tanzania. It is of paramount significance to acknowledge the overarching global strategies that have been meticulously tailored to fashion employment opportunities specifically for the youth, ensuring their active involvement and contribution to the agricultural sector, which stands as an invaluable pillar of the enchanting continent of Africa. Within this mesmerizing tapestry, the noble aim is to skillfully carve out pathways for the vibrant youths, empowering them to embark upon agriculture as an exhilarating, lucrative, and truly gratifying vocation. The call to action resonates with urgency and utmost sincerity as these remarkable nations urge the young generation wholeheartedly to seize the golden opportunity that gleams eagerly before them. For in the vibrant pulse of the youthful generation resides the transformative power and creative energy that has the potential to reshape the agricultural landscape of Africa for years to come. The diverse experiences encountered by these extraordinary nations may have differed subtly, as each country navigates unique challenges and triumphs in their agricultural endeavors. Yet, they were united by a multitude of commonalities in both their triumphs and their tribulations. One vital thread that binds these nations together is the pursuit and implementation of common policy enactments that have paved the way for progress. Each country persistently strives towards formulating and implementing an all-encompassing agricultural and natural resources education and training policy that resonates with their specific context and aspirations. This ambitious policy stems from an unyielding determination to diligently evaluate and invigorate education and training within the vast agriculture sector. These nations fervently endeavor to magnetize and retain a higher influx of African youths, enticing them towards agriculture and its associated disciplines, whilst offering them alluring prospects that render the field irresistible to pursue academically. The ultimate goal is to leave them equipped with indispensable qualifications and skills, enabling them to thrive in the agricultural landscape of the future. To successfully embark upon this ambitious journey, a succession of decisive steps has been undertaken, encompassing the indispensable development and reformulation of curricula entitled to secondary schools and higher education institutions. The commitment to brilliance is palpable, as specialists immersed in curriculum development, teaching, research, and extension have dedicated themselves with unwavering passion and dedication to ensuring a high standard of education in agriculture. In addition, unwavering emphasis has been placed upon capacity development in the vital domains of monitoring and evaluation within the national university and ministry staff. This focus aims to foster an environment of exploration and innovation, where knowledge serenades progress. By enhancing the skills and knowledge of staff members, these nations are building a strong foundation for sustainable growth in the agricultural sector, paving the way for greater achievements and advancements. A crucial element in the realization of these aspirations has manifested in traceable higher budget allocations to the education sector. This strategic allocation effectively renders education the recipient of abundant resources and staff amplification in schools and universities alike. The harmonious coalescence of sufficient resources, expertise, and manpower creates a landscape that underpins the destined growth and prosperity of African agriculture, empowering the youth to manifest their fullest potential. Diving headfirst into the vibrant endeavors of Mali, a resounding testament to their relentless pursuit emerged. In Mali, they have boldly adopted a sub-regional and regional approach to majestic curriculum reform, forging invaluable collaborations among the manifold faculties of agriculture and natural resources at the university level. This collaborative spirit extends not only within the boundaries of Mali but also reaches beyond national borders, engaging neighboring countries in the pursuit of knowledge and progress. Notably, the esteemed faculty of Les Ecoles Inter-Etats des Ingenieurs de Equipment Rural, a prestigious technical school, has proudly joined this transformative movement, gracing Burkina Faso, Niger, Mali, and Togo. The remarkable outcome of this concerted effort naturally materialized as the youthful generation in Mali boldly embraced the tantalizing realm of agriculture. Through the lens of a technical and scientific narrative, they embarked upon a journey of enlightenment that encompassed a truly formidable transformation of the agricultural sector. This transformative journey has ignited a passion for agriculture among young minds, equipping them with the skills and knowledge to contribute significantly to the development and growth of their nation's agriculture. The awe-inspiring policy in Rwanda has erected an exquisite framework teeming with enticing incentives, designed to captivate and nourish the hearts and minds of ambitious students. The policy resonates deeply with dedicated teachers who collectively embark upon the pursuit of knowledge and career success within the realms of science and agriculture. Rwanda's enlightened policy radiates in the form of enchanting scholarship programs, specifically tailored to embrace and uplift students on their arduous educational voyage. This ambitious educational framework in Rwanda is fortified by favorable advancements in the realm of salary increases lavished upon the esteemed teachers who have chosen to devote themselves wholeheartedly to this noble cause. Empowering and motivating teachers with improved remuneration not only recognizes their invaluable contribution but also creates a supportive environment that fosters excellence in education and cultivates a passion for agriculture among students. Alongside this, the provision of indispensable teaching materials to both students and teachers alike generates an ethereal ambiance within the educational landscape. An Aura of Enlightenment cascades effortlessly throughout the corridors of academia, beautifully engendering success and unleashing the full potential of Africa's youth. These teaching materials serve as beacons of knowledge, allowing students to immerse themselves in a world of discovery, innovation, and growth. As the curtain falls on this captivating review, it becomes abundantly clear that the efforts exerted by these remarkable nations of Ghana, Mali, Rwanda, Sao Tome & Principe, Senegal, and Tanzania are not in vain (Nji et al. 2022). The relentless pursuit of youth participation in the agricultural sector echoes with a resounding message of hope, perseverance, and unyielding determination. It stands as a testament to the unwavering belief in the untapped potential of Africa's youth and the immense value they bring to the realm of agriculture. Through collaborative policies, targeted investments, meticulous curriculum reform, and a deep commitment to knowledge dissemination, these nations are paving the way for a future where African agriculture thrives, and its youth take center stage as agents of change and progress. As the sun rises on the horizon of Africa's agricultural landscape, it illuminates a path of unparalleled opportunity, beckoning the vibrant youths to step forward and embrace the rewarding journey that lies ahead. Together, united in their pursuit, Africa's nations will continue to create a brighter, more sustainable future for generations to come (Akpuokwe et al. 2024; Rauf and Tariq 2024; Castro et al. 2021; Abdelfattah and El-Shamy 2024; Sujadmiko et al. 2023; Marder 2020; Siddiquee 2023; Saks 2022; Lin and Hong 2022; Nji et al. 2022; Gills and Hosseini 2022).

References

Abdelfattah I, El-Shamy AM (2024) Review on the escalating imperative of zero liquid discharge (ZLD) technology for sustainable water management and environmental resilience. J Environ Manag 352:120003. https://doi.org/10.1016/j.jenvman.2024.120003

Abdulai AR (2022) A new Green Revolution or neoliberal entrenchment in agri-food systems? Exploring narratives around digital agriculture, food systems, and development in Africa. J Dev Stud 58(11):2041–2058. https://doi.org/10.1080/00220388.2022.2048576

Abobi SM, Wolff M (2020) West African reservoirs and their fisheries: an assessment of harvest potential. Ecohydrol Hydrobiol 20(3):451–462. https://doi.org/10.1016/j.ecohyd.2020.03.002

Abu K (2023) The separateness of spouses: conjugal resources in an Ashanti town. In: Oppong C (ed) Female and male in West Africa. Routledge, pp 93–114

Adebayo WG (2023) Cassava production in Africa: a panel analysis of the drivers and trends. Heliyon 9(4):e15021. https://doi.org/10.1016/j.heliyon.2023.e15021

Adebayo WG, Silberberger M (2020) Poverty reduction, sustainable agricultural development, and the cassava value chain in Nigeria. In: Crush J, Caesar F (eds) The palgrave handbook of agricultural and rural development in Africa. Palgrave Macmillan, pp 525–551. https://doi.org/10.1007/978-3-030-41509-0_24

Agarwal B (2022) Women and technological change in agriculture: the Asian and African experience. In: Agarwal B (ed) Technology and Rural Women. Oxford University Press, pp. 15–42

Ahn J, Kim J, Kwon O (2022) Unified almost linear kernels for generalized covering and packing problems on nowhere dense classes. Proceedings of the ACM on Algorithms, 6(3). Article 45. https://doi.org/10.1145/3530804

Ainsworth R, Cowx IG, Funge-Smith SJ (2021) A review of major river basins and large lakes relevant to inland fisheries. FAO Fisheries and Aquaculture Technical Paper No. 608. Food and Agriculture Organization of the United Nations. https://www.fao.org

Akanmu AO, Olowe OM, Phiri AT, Nirere D, Odebode AJ, Karemera Umuhoza NJ, Babalola OO (2023) Bioresources in organic farming: implications for sustainable agricultural systems. Horticulturae 9(6):659. https://doi.org/10.3390/horticulturae9060659

Akpuokwe CU, Adeniyi AO, Bakare SS, Eneh NE (2024) Legislative responses to climate change: a global review of policies and their effectiveness. Int J Appl Res Soc Sci 6(3):225–239. https://doi.org/10.51594/ijarss.v6i3.783

Alajangi S, Prasad NSR, Baruah A (2021) Monitoring and changing pattern of shifting cultivation and reclamation in hilly regions using geospatial technology. In: Modern cartography series. Springer, pp 211–228. https://doi.org/10.1007/978-3-030-61950-4_14

Amusan L, Oyewole S (2023) Precision agriculture and the prospects of space strategy for food security in Africa. Afr J Science Technology Innov Dev 15(3):325–336. https://doi.org/10.1080/20421338.2022.2065751

Antwi-Agyei P, Dougill AJ, Doku-Marfo J, Abaidoo RC (2021) Understanding climate services for enhancing resilient agricultural systems in Anglophone West Africa: the case of Ghana. Climate Services 22:100215. https://doi.org/10.1016/j.cliser.2021.100215

Araujo GS, Silva JWAD, Cotas J, Pereira L (2022) Fish farming techniques: current situation and trends. J Mar Sci Eng 10(11):1598. https://doi.org/10.3390/jmse10111598

Asante Gyamerah S, Ikpe D (2021) A review of effects of climate change on Agriculture in Africa. [PDF]

Assistant JHET, BS Gaiya A (2023) How can I liberate the slaves? The Neglected Tradition of Developmental Abolitionism. osf.io

Ball JA (2020) Women farmers in developed countries: a literature review. Agriculture and Human Values. washburn.edu

Barthélemy ZIE, Yacouba BAMBA, Jean-Luc BK, Delphin KA, Adouony EN, Noel GROGA, Allassane O (2024) Evaluation of the impact of agricultural by-products on the integrated production of merchantable Tilapia Oreochromis Niloticus (Linné, 1758) and Rice (Oryza Glaberrima, Steud, 1853) in Bonoufla, Côte D'ivoire.

Beg MM, Roy SM, Moulick S, Mandal B, Kim T, Mal BC (2024) Economic feasibility study of organic and conventional fish farming systems of Indian major carps. Sci Rep 14(1):7001. https://www.nature.com

Bjornlund V, Bjornlund H, Van Rooyen AF (2020) Why agricultural production in sub-Saharan Africa remains low compared to the rest of the world–a historical perspective. Int J Water Resour Dev 36(suppl 1): S20–S53. https://www.tandfonline.com

Boafo J, Lyons K (2022) The rhetoric and farmers' lived realities of the Green Revolution in Africa: case study of the Brong Ahafo region in Ghana. J Asian Afr Stud [HTML]

Byerlee D, ynam JK (2020) The development of the international center model for agricultural research: a prehistory of the CGIAR. World Dev [HTML]

Byerlee D, Lynam KL (2020) The Invention of the International Agricultural Research Center Model and the CGIAR System. World Dev researchgate.net

Castellanza L (2019) Institutional contradictions and opportunities: How marginalized women farmers engage in bottom-up institution-building. osf.io
Castro GDR, Fernandez MCG, Colsa AU (2021) Unleashing the convergence amid digitalization and sustainability towards pursuing the Sustainable Development Goals (SDGs): a holistic review. J Clean Prod [HTML]
Ceballos H, Rojanaridpiched C, Phumichai C, Becerra LA, Kittipadakul P, Iglesias C, Gracen VE (2020) Excellence in cassava breeding: perspectives for the future. Crop Breed Genet Genom 2(2). hapres.com
Chang ZQ, Neori A, He YY, Li JT, Qiao L, Preston SI, Li J (2020) Development and current state of seawater shrimp farming, with an emphasis on integrated multi-trophic pond aquaculture farms, in China–a review. Rev Aquaculture 12(4):2544–2558. [HTML]
Christophe Jacques D, Defourny P (2019) Accuracy requirements for early estimation of crop production in Senegal. [PDF]
Clay N, Zimmerer KS (2020) Who is resilient in Africa's green revolution? Sustainable intensification and climate smart agriculture in Rwanda. Land use policy. sciencedirect.com
Cock JH, Connor DJ (2021) Cassava. Crop physiology case histories for major crops. cgiar.org
D Michler J, Josephson A, Kilic T, Murray S (2020) Estimating the Impact of Weather on Agriculture. [PDF]
Dahunsi SO, Fagbiele OO, Yusuf EO (2020) Bioenergy technologies adoption in Africa: a review of past and current status. J Clean Prod lmu.edu.ng
Danso-Abbeam G, Baiyegunhi LJS, Ojo TO (2020) Gender differentials in technical efficiency of Ghanaian cocoa farms. Heliyon 6(5):e04412. https://doi.org/10.1016/j.heliyon.2020.e04412
Dasgupta A, Johnson-Kanu A (2020) Precolonial states and development: evidence from Agriculture in Africa. OSF Preprints. https://doi.org/10.31219/osf.io/xxxxx
Djihad T, Alghorbany A, Izzuddin Bin Muhamad A, Alam MM (2021) Government policies, financial scopes and technological usages for agricultural development and post-harvest loss reduction in Algeria. OSF Preprints. https://doi.org/10.31219/osf.io/xxxxx
Duncan E, Abdulai AR, Fraser EDG (2021) Modernizing agriculture through digital technologies: prospects and challenges. In: Handbook on the human impact of agriculture. Edward Elgar Publishing, pp 138–161
Ema EOS, Obidiegwu JE, Chilaka CA, Akpabio EM (2023) Indigenous food yam cultivation and livelihood practices in Cross River State, Nigeria. World 4(1): 1–17. https://doi.org/10.3390/world4010001
Enisan Akinlo A, Michael Ojo S (2022) Economic consequences of the COVID-19 pandemic on sub-Saharan Africa: a historical perspective. OSF Preprints. https://doi.org/10.31219/osf.io/xxxxx
Fastellini G, Schillaci C (2020) Precision farming and IoT case studies across the world. In: Agricultural Internet of Things and decision support for precision smart farming. Academic Press, pp 331–415. https://doi.org/10.1016/B978-0-12-818373-1.00012-3
Fonjong LN, Gyapong AY (2021) Plantations, women, and food security in Africa: Interrogating the investment pathway towards zero hunger in Cameroon and Ghana. World Dev 138:105293. https://doi.org/10.1016/j.worlddev.2020.105293
Food and Agriculture Organization of the United Nations, African Union Commission, & Ministries of Agriculture. (2010–2023) Crop profiles and agro-ecological suitability of major food crops in Africa. https://www.fao.org
Francisco Ribeiro P, Camargo Rodriguez AV (2020) Emerging advanced technologies to mitigate the impact of climate change in Africa. Plants 9(12):1742. https://doi.org/10.3390/plants9121742
Gamo Longang S, Paul N, Tankeu Y (2023) Food security effects of agricultural land expansion in sub-Saharan Africa. OSF Preprints. https://doi.org/10.31219/osf.io/xxxxx
German LA, Bonanno AM, Foster LC, Cotula L (2020) "Inclusive business" in agriculture: Evidence from the evolution of agricultural value chains. World Dev 134:105018. https://doi.org/10.1016/j.worlddev.2020.105018
Giller KE (2020) The food security conundrum of sub-Saharan Africa. Glob Food Secur 26:100431. https://doi.org/10.1016/j.gfs.2020.100431
Giller KE, Delaune T, Silva JV, Descheemaeker K, van de Ven G, Schut AG et al (2021) The future of farming: who will produce our food? Food Secur 13(5):1073–1099. https://doi.org/10.1007/s12571-021-01184-6
Gills BK, Hosseini SAH (2022). Pluriversality and beyond: consolidating radical alternatives to (mal-)development as a commonist project. Sustain Scie 17:1461–1475. https://doi.org/10.1007/s11625-022-01118-5
Glazebrook T, Noll S, Opoku E (2020) Gender matters: Climate change, gender bias, and women's farming in the global South and North. Agriculture 10(7):267. https://doi.org/10.3390/agriculture10070267
Gokool S, Mahomed M, Kunz R, Clulow A, Sibanda M, Naiken V et al (2023) Crop monitoring in smallholder farms using unmanned aerial vehicles to facilitate precision agriculture practices: a scoping review and bibliometric analysis. Sustainability 15(4):3557. https://doi.org/10.3390/su15043557
Harwood J (2020) Could the adverse consequences of the green revolution have been foreseen? How experts responded to unwelcome evidence. Agroecol Sustain Food Syst 44(2):153–170. https://doi.org/10.1080/21683565.2019.1649634
Hasan K, Tanaka TST, Alam M, Ali R (2020) Impact of modern rice harvesting practices over traditional ones. Rev Agricultural Sci 8:1–12. https://doi.org/10.7831/ras.8.0_1
Hurt RD (2020) The green revolution in the Global South: science, politics, and unintended consequences. University of Arkansas Press
Ikuemonisan ES, Mafimisebi TE, Ajibefun I, Adenegan K (2020) Review of cassava production in Nigeria: trends and decomposition analysis approach. Tarım Ekonomisi Araştırmaları Derg 6(2):102–114

Inegbedion HE, Inegbedion EE, Obadiaru ED, Asaleye AJ, Ayeni A, Aremu C (2020) Cassava attractiveness in Nigeria: a policy improvement approach. J Agribus Dev Emerg Econ 10(2):157–175. https://doi.org/10.1108/JADEE-07-2018-0094

Jayasekara DN (2021) Can traditional farming practices explain attitudes towards scientific progress? Econ Model 96:222–235. https://doi.org/10.1016/j.econmod.2020.03.011

Jayne TS, Fox L, Fuglie K, Adelaja A (2021) Agricultural productivity growth, resilience, and economic transformation in sub-Saharan Africa. Association of Public and Land-grant Universities (APLU)

Jellason NP, Robinson EJZ, Ogbaga CC (2021) Agriculture 4.0: Is sub-Saharan Africa ready? Appl Sci. 11(2):575. https://doi.org/10.3390/app11020575

Kaluba Jickson C (2021) Food security and state agricultural policies: the long history of Cassava in Zambia from the pre-colonial period to 1990. Africa: Riv Semestrale di Stud e Ricerche: NS III(2):53–78

Kamagaté B, Ouattara N, Bi UZ, Pelebe R (2021) Socio-economic study of rice-fish farming in the region of upper Sassandra (Côte d'Ivoire). Moroccan J Agricultural Sci 2(1):1–9

Kansanga MM, Mkandawire P, Kuuire V, Luginaah I (2020) Agricultural mechanization, environmental degradation, and gendered livelihood implications in northern Ghana. Land Degrad & Dev 31(11):1422–1440. https://doi.org/10.1002/ldr.3536

Karunathilake E, Le AT, Heo S, Chung YS, Mansoor S (2023) The path to smart farming: innovations and opportunities in precision agriculture. Agriculture 13(2):392. https://doi.org/10.3390/agriculture13020392

Kirkwood D (2022) Settler wives in Southern Rhodesia: a case study. Routledge

Kruseman G (2022) Understanding the regional landscape and contributions of crops to end hunger in the Global South: towards a parsimonious foundation. OSF Preprints. https://doi.org/10.31219/osf.io/xxxxx

Kwaghtyo DK, Eke CI (2023) Smart farming prediction models for precision agriculture: a comprehensive survey. Artificial Intelligence Review 56:6173–6210. https://doi.org/10.1007/s10462-022-10298-4

Langyintuo A (2020) Smallholder farmers' access to inputs and finance in Africa. In: The role of smallholder farms in food and nutrition security. Springer, pp 133–152. https://doi.org/10.1007/978-3-030-42148-9_8

Lin L, Hong Y (2022) Developing a green bonds market: lessons from China. European Business Organization Law Review 23(2):285–316. https://doi.org/10.1007/s40804-021-00226-6

Lynam J, Byerlee D, Moock JL (2024) The organizational challenge of international agricultural research: The fifty-year odyssey of the CGIAR. Food Policy 122:102561. https://doi.org/10.1016/j.foodpol.2023.102561

Mapiye O, Makombe G, Molotsi A, Dzama K, Mapiye C (2021) Towards a revolutionized agricultural extension system for the sustainability of smallholder livestock production in developing countries: the potential role of digital technologies. Sustainability 13(11):5861. https://doi.org/10.3390/su13115861

Marder ID (2020) The new international restorative justice framework: Reviewing three years of progress and efforts to promote access to services and cultural change. Int J Restorative Just 3(3):395–414. https://doi.org/10.5553/IJRJ.000035

Munguti JM, Nairuti R, Iteba JO, Obiero KO, Kyule D, Opiyo MA et al (2022) Nile tilapia (Oreochromis niloticus Linnaeus, 1758) culture in Kenya: Emerging production technologies and socio-economic impacts on local livelihoods. Aquaculture Fish Fish 2(4):265–276. https://doi.org/10.1002/aff2.47

Musa S (2023) Fishing for answers: should afrotropical farmers use cages or ponds for grow-out of Nile tilapia? In: Emerging sustainable aquaculture innovations in Africa. Springer Nature, pp 411–426. https://doi.org/10.1007/978-981-19-0306-5_18

Nemade S, Ninama J, Kumar S, Pandarinathan S, Azam K, Singh B, Ratnam KM (2023) Advancements in agronomic practices for sustainable crop production: a review. Int J Plant Soil Sci 35(22):679–689. https://doi.org/10.9734/ijpss/2023/v35i2232150

Neupert-Wentz C, Müller-Crepon C (2021) Traditional institutions in Africa, past and present. OSF Preprints. https://doi.org/10.31219/osf.io/xxxxx

Nivievskyi O, Martyshev P, Kvasha S (2023) Agricultural policy in Ukraine. Kyiv School of Economics

Nji QN, Babalola OO, Mwanza M (2022) Aflatoxins in maize: can their occurrence be effectively managed in Africa in the face of climate change and food insecurity? Toxins 14(5):320. https://doi.org/10.3390/toxins14050320

O Achieng AB, Arhonditsis GE, Mandrack NM, Febria C, Opaa BJ, Coffey T, Irvine KO, Masese FM, Ajode ZE, Barasa J, Obiero K, Kaunda-Arara B (2023) Monitoring biodiversity loss in rapidly changing Afrotropical ecosystems: an emerging imperative for governance and research. [PDF]

Ochieng J, Afari-Sefa V, Muthoni F, Kansiime M, Hoeschle-Zeledon I, Bekunda M, Thomas D (2022) Adoption of sustainable agricultural technologies for vegetable production in rural Tanzania: Trade-offs, complementarities and diffusion. Int J Agric Sustain 20(4):478–496. https://doi.org/10.1080/14735903.2021.1959142

Odekina FO (2023) Trend analysis of Cassava Production in Nigeria and Thailand. International Journal of Global Affairs Res Dev 1(1):117–125

Ogura S, Forwell SJ, Backman C, Lincoln NK, Bulkan J, Odakawa K et al (2024) Meaning of small grains cultivation as a traditional occupation in Japan. J Occupat Sci 1–16. https://doi.org/10.1080/14427591.2024.2324567

Onyiriuba L, Okoro EUO, Ibe GI (2020) Strategic government policies on agricultural financing in African emerging markets. Agric Finance Rev 80(4):563–586. https://doi.org/10.1108/AFR-01-2020-0007

Oppong C (2023) Female and male in West Africa. Routledge

Pierotti RS, Friedson-Ridenour S, Olayiwola O (2022) Women farm what they can manage: how time constraints affect the quantity and quality of labor for married women's agricultural production. World Dev. 153:105829. https://doi.org/10.1016/j.worlddev.2022.105829

Pingali P (2007) Agricultural growth and economic development: A view through the globalization lens. Agricultural Economics 37(s1):1–12. https://doi.org/10.1111/j.1574-0862.2007.00231.x

Plucknett DL (2020) International goals and the role of the international agricultural research centers. In: Sustain Agric Syst. CRC Press, pp 97–118

Polanco A, Newman MEJ (2023) Hierarchical core-periphery structure in networks. OSF Preprints. https://doi.org/10.31219/osf.io/xxxxx

Rai AK, Kumar N, Katiyar D, Singh O, Sreekumar G, Verma P (2023) Unlocking productivity potential: the promising role of agricultural robots in enhancing farming efficiency. Int J Plant Soil Sci 35(18):624–633. https://doi.org/10.9734/ijpss/2023/v35i1831476

Raimi L, Panait M, Sule R (2021) Leveraging precision agriculture for sustainable food security in sub-Saharan Africa: a theoretical discourse. In: Shifting patterns of agricultural trade: the protectionism outbreak and food Security. Springer, pp 491–509. https://doi.org/10.1007/978-3-030-36822-7_22

Raj M, Gupta S, Chamola V, Elhence A, Garg T, Atiquzzaman M, Niyato D (2021) A survey on the role of Internet of Things for adopting and promoting Agriculture 4.0. J Netw Computer Appl 187:103107. https://doi.org/10.1016/j.jnca.2021.103107

Rauf S, Tariq M (2024) The bilateral engagement of India and Pakistan at Shanghai cooperation organization: prospects and challenges. NUST J Int Peace Stab 7(1):1–17. nust.edu.pk

Ricome A, Louhichi K, Gomez y Paloma S (2023) Subsidizing agricultural inputs in Senegal: comparative analysis of three modes of intervention using a farm household model. Joint Research Centre, European Commission.

Rohne Till E (2021) A green revolution in sub-Saharan Africa? The transformation of Ethiopia's agricultural sector. J Int Dev 33(6):1031–1048. https://doi.org/10.1002/jid.3562

Sackey Teye E, Tetteh Quarshie P (2021) Impact of agriculture finance in modern technologies adoption for enhanced productivity and rural household economic wellbeing in Ghana: a case study of rice farmers in Shai-Osudoku District. OSF Preprints. https://doi.org/10.31219/osf.io/xxxxx

Saks M (2022) Responsible leadership: essential to the achievement of the UN Sustainable Development Goals. Sustainability 14(3):1385. https://doi.org/10.3390/su14031385

Santpoort R (2020) The drivers of maize area expansion in sub-Saharan Africa. How policies to boost maize production overlook the interests of smallholder farmers. Land 9(9):300. https://doi.org/10.3390/land9090300

Schildkrout E (2023) Dependence and autonomy: the economic activities of secluded Hausa women in Kano. In: Oppong C (ed) Female and male in West Africa. Routledge, pp xx–xx

Scott GJ (2021) A review of root, tuber and banana crops in developing countries: past, present and future. Int J Food Sci Technol 56(3):1093–1111. https://doi.org/10.1111/ijfs.14814

Seidensticker D, Jungnickel KVM (2023) Early food production in the Congo Basin. OSF Preprints. https://doi.org/10.31219/osf.io/xxxxx

Sekaran U, Lai L, Ussiri DAN, Kumar S, Clay S (2021) Role of integrated crop-livestock systems in improving agriculture production and addressing food security–A review. J Agric Food Res 5:100190. https://doi.org/10.1016/j.jafr.2021.100190

Seranne M, Anka Z (2005). South Atlantic continental margins of Africa: a comparison of the tectonic vs climate interplay on the evolution of equatorial West Africa and southwest Africa margins. Geological Society Special Publication

Shaikh TA, Rasool T, Lone FR (2022) Towards leveraging the role of machine learning and artificial intelligence in precision agriculture and smart farming. Comput Electron Agric 198:107119. https://doi.org/10.1016/j.compag.2022.107119

Siddiquee NA (2023) Right to information law and open government: the Indonesian experience. In: Open government and freedom of information: policy and practice in Asia and the Middle East. Routledge, pp 185–203

Sinhababu DP, Poonam A (2022) Development in rice-fish-livestock farming for higher production and income in coastal areas. In: Transforming coastal zone for sustainable food and income security. Springer, pp 261–276. https://doi.org/10.1007/978-3-030-89736-9_16

Smidt HJ, Jokonya O (2022) Factors affecting digital technology adoption by small-scale farmers in agriculture value chains (AVCs) in South Africa. Inf Technol Dev 28(4):846–865. https://doi.org/10.1080/02681102.2021.1974011

Spindelman D, Crouch LA (2024) Purpose of education in former British African colonies: from contestation to bureaucratization. OSF Preprints. https://doi.org/10.31219/osf.io/xxxxx

Stapleton SC (2022) More than the sum of its parts: the Consultative Group on International Agricultural Research, CGIAR. J Agric Food Inf 23(2):93–110. https://doi.org/10.1080/10496505.2022.2033629

Sujadmiko B, Narazua Y, Natamihardja R, Meutia IF (2023) ASEAN challenges toward supranational organization. Russian Law J 11(5S):238–248

Translation & Literary Studies AWEJ, Hamad W (2021) Exploring the postcolonial concept through the eye of European expansionism and imperialism. OSF Preprints. https://doi.org/10.31219/osf.io/xxxxx

Udoumoh UI, Ikrang EG, Ehiomogue PO (2021) Precision farming and fertilizer recommendation using geographic information system (GIS): a review. Int J Agricult Earth Sci 7(3):45–58

Walther O, Retaille D (2019) Mapping the Sahelian space. OECD Publishing

Watts N, Scales IR (2020) Social impact investing, agriculture, and the financialisation of development: Insights from sub-Saharan Africa. World Dev 130:104889. https://doi.org/10.1016/j.worlddev.2020.104889

Yin L, Ghosh R, Lin C, Hale D, Weigl C, Obarowski J, Zhou J, Till J, Jia X, Mao T, Kumar V, Jin Z (2023) Mapping smallholder cashew plantations to inform sustainable tree crop expansion in Benin. OSF Preprints. https://doi.org/10.31219/osf.io/xxxxx

4 The Dual Challenge: Conservation and Food Security in Africa

4.1 The Nexus Between Conservation and Food Security

Developing and industrializing countries are likely to hold the answer to the dual challenge that the world faces in terms of food production and conservation. A substantial increase in local food production may be achieved through import substitution, a financially attractive approach driven by powerful forces. However, due to the extent and type of land and water resources, as well as the diversity of ecosystems in developing countries, the pursuit of both conservation and expanded food production can be a delicate balancing act. It requires carefully navigating a narrow path on a sharp ridge. Straying too far towards conservation may intensify food-related issues, while leaning too heavily towards expanded production may result in the depletion of vital resources. Unfortunately, the recognition of this tightrope situation often comes too late, after irreversible damage has already occurred in the form of landslides. Extensive methodological and theoretical developments, as well as pilot studies, have revealed the complexities inherent in addressing this challenge. These complexities emphasize the need for close collaboration and two-way interaction between conservation scientists and production specialists. In order to integrate conservation features into production projects effectively, it is crucial to have a predictive understanding of how specific activities impact natural resources, coupled with the ability to monitor these impacts. As people have become increasingly aware of the challenges and scarcities involved in managing resources effectively, there has been a growing interest in the development and application of analytical methods for decision-making. This interest has been particularly notable in the United States and the Food and Agriculture Organization (FAO). These analytical methods encompass a wide range of approaches, all aimed at improving the allocation of resources among various possible uses. By utilizing these methods, countries can make more informed decisions regarding resource management, ultimately contributing to sustainable development on a global scale. Developing and industrializing countries hold immense potential and offer viable solutions to the pressing global issues of food production and conservation. Achieving significant growth in local food output can be realized through import substitution, an economically favorable approach propelled by influential factors. However, owing to the vastness and heterogeneity of land and water resources, as well as the diverse ecosystems present in developing nations, attaining a balance between conservation and expanded food production is a delicate undertaking. It necessitates skillfully navigating a narrow path along a steep ridge, where veering too far towards conservation may exacerbate food-related challenges, while leaning excessively towards increased production may lead to the

E. D. Wiafe, *Wildlife Conservation and Agriculture Development in Africa*, Sustainable Development Goals Series, https://doi.org/10.1007/978-3-032-09233-5_4

depletion of vital resources. Tragically, acknowledgement of this high-wire situation often arrives too belatedly, after irreversible damage manifests in the form of devastating landslides. Comprehensive methodological and theoretical advancements, alongside pilot studies, have uncovered the intricate complexities inherent in tackling this challenge. These intricacies underscore the imperative for close collaboration and bi-directional synergy between conservation scientists and production specialists. To seamlessly integrate conservation elements into production endeavors, a prescient comprehension of how individual activities affect natural resources, combined with the capacity to vigilantly monitor these impacts, is paramount. These analytical methods encompass a broad spectrum of approaches, all aimed at optimizing the allocation of resources across diverse potential uses. By harnessing the power of these methods, nations can make well-informed decisions concerning resource management, thereby contributing to sustainable global development in a profound manner (Amelework et al 2021; Kurbatova et al. 2020; Ragasa et al. 2020; Amirova et al. 2020; Fastovich and Kapsargina 2020; Nusratovich and Asidaxon 2023; Naglis-Liepa et al. 2021; Kuiper and Cui 2021a, b; Irwin 2021).

One of the most influential and recurring themes in the last half-century of African development studies is the interconnectedness of conservation and development. Tomorrow's countryside vision for Africa cuts a sharp contrast with the recent past and present. Recurrent food shortages combined with agricultural modernization policies raise the specter of a regional "population squeeze" in which the numbers of poor people spiral upwards as resources per capita decline. The global significance of this phenomenon, concerns about Africa's relative isolation from global food and commodity markets, and widespread dissatisfaction with past structural adjustment policies have recently led to calls for a "second generation" of economic reforms for Africa, including increased support for agricultural development. These all seem likely to combine with continued rapid population growth and urbanization to greatly increase the colonization of Africa's remaining wildlands, with potential impacts on wildlife and the prospects for long-term biodiversity conservation that varies from location to location but is everywhere of great significance. The late Prof David Kimble, an eminent agricultural and development economist and scholar of Africa, once commented that "the environmental crises of the world today are the product of poverty and the crises of the future will be the product of development". It is hoped, of course, that the future will be different from the scenario he envisaged, but the road to sustainable growth and development may well be a tortuous one, with many pitfalls and switchbacks. The challenges that lie ahead are immense, as Africa strives to balance the need for economic progress with the imperative of environmental conservation. Additionally, finding a delicate equilibrium between the aspirations of its people and preserving its unique biodiversity becomes a paramount task. The crucial and complex long-term interactions between food security and biodiversity have become a major focus of policy and research over the last two decades. Studies both global and local show that the indirect use values of free goods and ecosystem services are often many times the direct use values of marketed products. This is as true for the rural poor of Africa as it is for the global human population in relation to climate regulation or the maintenance of marine fisheries. Nonetheless, rural people's personal environmental ethics are quite compatible with "stated preference" methods used to value non-use goods. Rural Africans characteristically express patient discount rates, valuing a secure future for their children and grandchildren, and a widespread but not all-encompassing "inter-temporal" rather than "static" welfare maximand. This is reflected in decisions to conserve or plant trees, or to protect or clear woodlands and wetlands, often involving complex non-market opportunity costs. A key task for researchers here is to grasp the ethics and values in their own terms and not only as a residual of money-metric utility. A better understanding of these cultural land ethic traditions can inform conservation policies that seek to reconcile food security and poverty alleviation with environmental goals, rather than continually casting them in conflict. By embracing the principles of sustainable growth and development, Africa can chart a path towards a brighter future where both the well-being of its people and the protection of its natural resources are upheld. It is a journey that demands careful navigation, strategic decision-making, and collaborative efforts. Only by recognizing the intricate relationship between conservation and development can Africa truly achieve its potential and safeguard its invaluable biodiversity for generations to come. Africa stands at a critical juncture, poised to redirect its course by embracing innovative strategies that harmonize economic growth with environmental stewardship. The transformative power of such an approach cannot be underestimated. Through comprehensive planning, effective governance, and widespread engagement, Africa can navigate the challenges ahead, forging a new paradigm that ensures sustainable prosperity and enduring ecological health. The stakes are high, but the opportunities are abundant. Ultimately, the destiny of Africa lies in the hands of its visionary leaders and dedicated advocates who will shape a legacy of thriving ecosystems and equitable societies. Let us unite in the pursuit of a future where Africa's natural treasures are passionately conserved, its people prosper, and its legacy shines as a beacon of hope for the world (Ngwu and Onah 2023; Ramutsindela and Mickler 2020; Malemba et al. 2021; Mpofu et al. 2023; Lee and Gambiza 2022; Tamasiga et al. 2023; Falayi et al. 2021a, b; Asani 2020; Krauss 2021a, b; Kumar and Giri 2020a, b).

4.2 Conservation Efforts in Africa

Given the immense diversity of resource landscapes and the wide array of livelihood strategies found across the vast continent of Africa, modern conservation has developed a rich tapestry of approaches that aim to address the complex challenges at hand. Broadly speaking, these strategies can be classified into two distinct categories: those that actively promote development and entail transforming human behavior to reduce resource consumption, and those that prioritize protection and the preservation of natural ecosystems. The promotion of development rests on the fundamental notion that by enhancing human well-being, alleviating poverty, and offering viable economic alternatives, conservation goals can be achieved. This perspective emphasizes the critical role played by humans in environmental degradation and endeavors to instigate behavioral changes by fostering a deep comprehension of the costs and benefits associated with resource utilization. Ultimately, this paradigm aims to alleviate the burden placed on natural systems by humans, thus ensuring their long-term sustainability. On the other end of the spectrum, there lie the strategies of protected area conservation and in situ preservation. Protected area conservation operates on the premise that habitats and species can be safeguarded against impending threats if human activities that pose risks are effectively eliminated. In essence, it centers its efforts on nurturing the well-being of species and preserving vital sanctuaries. In sharp contrast, in situ preservation seeks to uphold the integrity of genetic resources and curtail species loss by diligently maintaining and rehabilitating viable wild populations of indigenous flora and fauna within their native habitats. This approach stems from the belief that evolution is an ever-unfolding process, thus making it possible to regenerate ecosystems even from severely degraded states. By adopting this multifaceted palette of conservation methodologies, it is hoped that Africa's extraordinary natural heritage can endure and thrive. These endeavors, collectively informed by scientific insights and local wisdom, strive to strike a harmonious balance between human development and environmental preservation. Through such concerted efforts, we can pave the way for a sustainable future where ecosystems flourish, species thrive, and the intricate web of life is preserved for generations to come. The success of these efforts lies not only in their implementation but also in the continuous monitoring and adaptation of strategies to evolving circumstances. It is essential to recognize that conservation is not a static endeavor but rather a dynamic process that necessitates ongoing innovation, collaboration, and engagement with local communities. To further enhance the efficacy of conservation, it is imperative to foster international cooperation and coordination. Given the interconnectedness of ecosystems and the far-reaching impacts of human activities, addressing environmental challenges in Africa requires a global perspective and collective action. This entails sharing best practices, exchanging scientific knowledge, and harmonizing efforts to safeguard shared resources. Joint initiatives that transcend borders and address transboundary conservation issues can play a transformative role in securing the future of Africa's biodiversity and its critical ecological services. Education and awareness also assume paramount importance in the realm of conservation. By promoting environmental literacy, cultivating a sense of stewardship, and empowering communities with the tools and knowledge necessary to make informed decisions, conservation efforts can gain significant momentum. Education fosters a deeper appreciation for the interconnectedness of natural systems and nurtures a sense of responsibility towards their preservation. In this regard, the engagement of youth and the incorporation of traditional knowledge systems in educational programs serve as powerful agents of change, ensuring the continuity of conservation efforts in the generations to come. Moreover, the integration of technology and innovation can greatly augment the effectiveness of conservation initiatives. Cutting-edge tools such as remote sensing, geographic information systems, and data analysis provide valuable insights into ecosystem dynamics, habitat distribution, and species populations. This information can inform evidence-based decision-making, facilitate targeted interventions, and enable adaptive management strategies. Harnessing the power of technology not only enhances the efficiency and precision of conservation efforts but also opens new avenues for collaboration and knowledge sharing in the global conservation community. In conclusion, the conservation of Africa's natural heritage is an intricate and multifaceted endeavor. It necessitates a comprehensive approach that embraces sustainable development, habitat protection, species preservation, international cooperation, education, and technological innovation. By weaving together these diverse threads, we can aspire to safeguard Africa's extraordinary biodiversity, foster the harmonious coexistence of humans and nature, and ensure a sustainable future for the generations to come. Through our collective efforts and unwavering commitment, we can leave a lasting legacy of environmental stewardship and preserve the wonders of Africa's natural landscapes for eternity. (Tian et al. 2022).

4.2.1 Protected Areas

However, in common with the rest of the world, the ideals set out in the earlier half of the twentieth century are often compromised in the current social and economic climate. This compromise has never been more under threat than in Africa under the pressures of chronic underdevelopment and a rapidly burgeoning rural population. Reasons cited for the widespread failure of protected areas in Africa revolve around the perceived notion that such areas exclude human activity and human residents. This was an early ideal of protected areas which it is now realized is both impractical and unethical. In Kenya, it has been noted that the average wildlife reserve takes up to 54 years to reach full implementation and gazettement as a protected area. This long bureaucratic process has meant that local people and external immigrants have become settled in these areas and often wary of too far retribution. With most of these people living at or below subsistence level, the potential loss in human welfare from further extension or strict implementation of protected area legislation is considerable. High rural population densities in many countries also mean that it is difficult to find large tracts of uninhabited land. The loss in land available for agriculture and animal husbandry to an increasingly territorially confident wildlife community is seen by many rural Africans as a direct threat to food security. Human–wildlife conflict has been on the increase in many areas, and the severity of its impacts on local people result in increased antagonism towards the conservation movement in general. A culture of wildlife tolerance and positive conservation attitudes is hard to maintain when rural people view wildlife as a competitor and a threat to their standard of living. Globalization is also partly to blame for the current trends in land use change within and around protected areas. The movement of liberal economic ideologies and free trade policies has led many African nations to open their previously closed markets to the outside world. Both national and international investors are now rapidly buying up land and establishing agricultural enterprises in proximity to protected areas. Competitive land values and soaring prices for arable land in nearby areas have resulted in heightened political and economic pressures to de-gazette protected areas and or deplete their natural resources. This, in turn, threatens the delicate balance between conserving wildlife and addressing the needs and aspirations of local communities. The urgency to find sustainable solutions that can reconcile the interests of both wildlife conservation and human development has never been greater. It requires a paradigm shift in the approach towards protected areas in Africa, taking into account the complexities and challenges posed by socio-economic factors, political dynamics, and the realities of a changing world. It demands innovative strategies that can harmonize conservation efforts with poverty alleviation, sustainable livelihoods, food security, and inclusive participation of local communities. Only a multifaceted and holistic approach can pave the way for a new era of protected areas management, where not only wildlife thrives but also human well-being is uplifted. This requires integrated planning, strong governance, stakeholder engagement, capacity-building, empowerment, and equitable distribution of benefits derived from protected areas. By adopting such an approach, Africa can bring about a transformation, where protected areas become catalysts for socio-economic development, champions of biodiversity conservation, and beacons of hope for a sustainable future. This would result in a shift towards a more ecologically harmonious coexistence between humans and wildlife, fostering understanding, appreciation, and long-term collaboration for the benefit of all (Ojiambo 2022; Harmsen 2021; Gathuku 2022; Kalvelage et al. 2021; Kuntai 2020; Kameri-Mbote et al. 2021; Togoch 2020; Juma et al. 2020a, b; Gibbon 2021a, b; Amara et al. 2020).

4.2.2 Wildlife Conservation and Environmental Preservation

Wildlife conservation currently comes second to agriculture as a national land-use priority in Africa, on the assumption that it cannot support people, thus is valueless. The huge and ever-growing human population in Africa describes that the majority of Africans are rural poor, reliant on agriculture for their survival. In addition, Africa's present economic plight and development objectives create a situation in which conservationists have to compete for a share of diminishing funds and natural resources.

Throughout the world, there are more than 100,000 protected areas covering 12.2 % of the Earth's land surface, as well as hundreds of international wildlife management and conservation projects, all with the intent to conserve species and ecosystems. 29 percent of Africa's land has been zoned specifically for wildlife conservation and tourism, in an attempt to conserve the continent's extraordinary wildlife heritage and the remaining representative samples of its major habitat types. This, however, is loosely termed, and in many cases involves no more than protection from poaching and uncontrolled burning.

Wildlife does have value, which can be harvested in both consumptive and non-consumptive ways. In the case of wildlife tourism, the value of fauna and flora is left standing, encouraging investors/landowners to protect their natural assets. Research into the economic value of wildlife for rural communities is being increasingly necessitated in Africa, the aim being to improve both the quantity and quality of returns from wildlife, in comparison to other land uses (Turner et al. 2022; Cisneros-Araujo et al. 2021; Mashapa et al. 2021; Tyrrell et al. 2020; Reaser et al. 2021; Ntukey et al. 2022; Lindsey et al. 2020; Gargallo 2020a, b; Kariuki et al. 2021; Manyani 2020).

4.2.3 Environmental Policies

One of the major environmental policy shifts in Sub-Saharan Africa since the 1980s has been a significant transition from state command and control natural resource management to the devolution of user and ownership rights over these valuable resources. This transformation was necessary as the decision-making power was previously concentrated in centralized states, which often resulted in the degradation and mismanagement of resources due to the allocation of labor and capital to other sectors of the economy. During that period, national leaders, driven by foreign exchange needs and the repayment of debt, prioritized resource exploitation with little consideration for the environment.

Fortunately, current thinking has evolved, recognizing the importance of devolving resource rights to local communities. This shift empowers communities to make vital decisions regarding the management of their resources, allowing them to allocate labor and capital resources in a more sustainable manner. Consequently, they directly benefit from the preservation and protection of these valuable resources. While the devolution of resource rights generally leads to positive outcomes, it is essential to contemplate various factors that suggest that merely granting local communities rights may not guarantee optimal functioning of resource management institutions.

Over time, it has become evident that various community elites tend to exert their influence and exploit the devolved rights, resulting in the exclusion of marginalized and poorer members within their communities. This exclusionary behavior undermines the principles of equity and sustainability that underpin the devolution of resource rights. As such, it is crucial to recognize the importance of implementing institutional changes that support and safeguard the interests of all community members.

These institutional changes should enable the participation and representation of minority ethnic or class groups within these communities. By ensuring the inclusion of these groups, decision-making processes can become more democratic and reflective of the diverse needs and perspectives within the community. Moreover, it is imperative to establish external monitoring mechanisms to hold accountable those who may choose to exploit their communities' resources for personal gain. Such monitoring mechanisms can help detect and address instances of corruption or unfair practices, ensuring that the devolution of resource rights truly benefits the community as a whole.

By taking both devolution of resource rights and the implementation of institutional changes into account, Sub-Saharan Africa can create a more equitable and sustainable model of resource management. In doing so, local communities will not only benefit from safeguarding their valuable resources for future generations but also ensure that all members, regardless of their social status, can actively participate in and reap the rewards of resource management decisions. This comprehensive approach strives to strike a delicate balance between local autonomy, community development, and safeguarding natural resources for the collective well-being and prosperity of present and future generations (Adam et al. 2021; Tantoh; Ullah and Kim 2021; Adeyanju et al. 2021; Occhiali 2023; Kenjio 2020; Bruns 2021; Kohnert 2024a, b; Ntumva 2022; Joshi et al. 2021a, b).

4.2.4 Sustainable Agriculture

Sustainable agriculture is a method of practicing environmentally-friendly farming that aims to protect and preserve the planet's resources. It involves growing crops and raising livestock in a more ecologically and sustainable manner, taking into consideration the long-term impact on the environment.

This holistic approach to farming focuses on reducing the use of chemical fertilizers and pesticides, as well as implementing methods of farming that are less destructive to the environment. By doing so, sustainable agriculture aims to minimize the contamination of soil, water, and air, and to preserve the natural ecosystem.

The ultimate goal of sustainable agriculture is to produce food that can be consumed by both human beings and animals without compromising the health of future generations. This means finding ways to increase yields and improve crop quality, while simultaneously protecting the land's fertility and the overall ecosystem.

There are numerous methods of sustainable agriculture being practiced worldwide, each adapted to suit specific geographical and cultural contexts. In Sub-Saharan Africa, where food security is a major concern, sustainable agriculture is seen as a crucial strategy to alleviate hunger and enhance agricultural productivity.

One prominent method of sustainable agriculture is Conservation Farming, which revolves around the adoption of resource-saving technologies and farming systems. Through implementing soil-enhancing practices and crop production techniques, Conservation Farming aims to increase long-term returns from the land.

Unlike traditional farming approaches that prioritize short-term gains without considering the broader environmental and social costs, Conservation Farming seeks a more sustainable and profitable approach. By conserving the environment and promoting responsible farming practices, it addresses the dual challenge of ecological preservation and food security.

Promoting sustainable agriculture, such as Conservation Farming, is crucial not only for current food production but also for the well-being of future generations. By embracing these methods, we can ensure that agriculture remains a vital source of sustenance while safeguarding the health of our planet (Amede et al. 2023; Montt and Luu 2020; Corbeels et al. 2020; Rodenburg et al. 2021; Kuyah et al. 2021a, b; Assefa et al. 2020; Vanlauwe and Dobermann 2020; Balehegn et al. 2022; Chinseu et al. 2022; Shrestha et al. 2020).

4.2.5 Food Security Challenges in Africa

In 1996, the World Food Summit defined food security as existing "when all people at all times have access to sufficient, safe, nutritious food to maintain a healthy and active life." It is widely acknowledged and accepted that a household is considered to be food secure when it possesses a consistent and dependable access to an ample quantity of highly nutritious food. However, despite the fact that the African continent possesses more than 50% of the world's underutilized arable land, it unfortunately struggles to produce an adequate supply of food to sustain its rapidly growing population. Africa finds itself confronted with three crucial and interconnected challenges in terms of food security, namely food availability, food access, and food utilization.

When it comes to food availability, numerous factors hinder the situation. These encompass various environmental trends such as the persistent issue of land degradation, which directly impacts the lives of approximately 180 million people in Africa. This is further compounded by the alarming rate at which soil fertility is currently diminishing due to unsustainable agricultural practices like shifting cultivation and overgrazing. Additionally, the continent is confronted with the looming threat of water scarcity, with projections indicating that water resources will experience severe stress within the next 25 years. Furthermore, exacerbating the situation are the unexpected shocks caused by conflicts and their aftermath, which result in the displacement of population and the creation of large numbers of refugees and internally displaced persons. Moreover, the economic policies, particularly the structural adjustment programs imposed by the International Monetary Fund (IMF) and the World Bank, have a significant impact on the overall level of food production in Africa. Consequently, if the current trends persist, it is projected that Africa will transform into a net importer of food by the year 2020, despite the marginal improvement observed in food aid over the past decade. To put this into perspective, the level of food aid provided in 1996 stood at a relatively modest 6 million tons, which pales greatly in comparison to the levels in the 1970s and 1980s, where food aid amounted to approximately 12 million tons (Mugiyo et al. 2021; Okigbo and Anyaegbu 2021; Imathiu 2021a, b; Soumare et al. 2022; Mugiyo et al. 2021; Muhammad et al. 2020; Paliwal et al. 2021; Aderibigbe et al. 2022; Hossain et al. 2021; Popoola et al. 2022).

4.2.6 Limited Access to Resources

Non-farm income, especially from enterprises and work opportunities within villages and local towns, has been widely recognized as a significant and promising potential pathway out of poverty for rural households across the African continent. This valuable source of income serves as a vital lifeline, offering rural communities and individuals a means to improve their livelihoods, enhance their standard of living, and break free from the shackles of poverty.

In the context of rural Africa, it is crucial to acknowledge that the majority of households rely heavily on common property regimes as their primary means of accessing natural resources. These common property regimes, which encompass communal lands, forests, water bodies, and other shared resources, play a pivotal role in sustaining the livelihoods of rural communities. However, it is essential to shed light on the inherent limitations and challenges associated with such systems of resource distribution and access.

Common property resource systems, while providing a certain level of collective ownership and control over resources, are often characterized by restrictions and constraints that pose barriers to equitable access and utilization. These restrictions are manifested through customary norms, traditional practices, and social arrangements that govern resource allocation within the community. As a result, the extent of private ownership and control over specific resources is relatively limited in rural Africa.

One key aspect that should be considered within the discourse of common property systems is the issue of access limitations. These restrictions can manifest in various forms, including but not limited to, restricted access based on gender, ethnicity, social status, or membership in certain community groups. These exclusionary practices can further perpetuate existing inequalities and hinder the potential benefits that non-farm income generation could bring to disadvantaged rural households.

Recognizing and addressing these access limitations is crucial for promoting inclusive economic development and poverty reduction in rural Africa. Efforts should be made to ensure that non-farm income opportunities are accessible to all members of the community, regardless of their background or social standing. This requires implementing targeted policies and interventions that aim to dismantle discriminatory practices and create an enabling environment for equal participation and resource utilization.

The non-farm income presents a significant avenue for rural households to escape poverty in Africa. However, it is imperative to acknowledge the access limitations inherent in common property resource systems, which predominantly govern resource distribution in rural communities. By understanding and addressing these limitations, the full potential of non-farm income can be harnessed, leading to sustainable poverty alleviation and inclusive development for rural African communities (Neglo et al. 2021; Michailidis and Lazaridou 2020; Kehinde and Ogundeji 2023; Giller et al. 2021; Adeyonu et al. 2024; Neglo et al. 2021; Waarts et al. 2021; Lawi et al. 2022; Bezu 2021; Séogo, 2023).

A common property regime is a type of ownership system characterized by the existence of a resource that is non-excludable, meaning that people or groups within a society cannot prevent others from utilizing the resource. Moreover, the utilization of this resource leads to a reduction of its availability for others, which is commonly referred to as rivalry. The limited accessibility to the resource can manifest in various ways, such as exclusion from the resource itself or restrictions on usage. Nevertheless, it is more often the case that the limitation pertains to the extent or severity of the aforementioned conditions, ensuring that the resource is accessible to all individuals, yet its usage is carefully regulated to maintain its long-term sustainability, promote environmental conservation, and ensure an equitable distribution among all its users, thereby fostering harmony and cooperation within the society in question. By implementing mechanisms such as community-based rules and regulations, participatory decision-making processes, and collective management practices, common property regimes strive to strike a balance between individual needs and the collective interest, creating a framework that encourages active engagement, responsibility, and stewardship. In this way, the common property regime acts as a catalyst for social cohesion, promoting inclusivity, cooperation, and the sustainable use of shared resources for the benefit of present and future generations (Kassis 2023; Slaev and Daskalova 2020; Abubakari et al. 2023; Salustri 2021; Gümplová 2023a, b; Casprini et al. 2023; Litsardou and Klabatsea; Wu 2022; Rayamajhee and Paniagua 2021; Wolsink 2020).

This then leads to a gradual deterioration and decline in the commonly shared pool resource by those individuals who are able to gain access to it, which subsequently results in a significant and adverse impact on the long-term sustainability as well as the future availability of the resource for all users. A concrete and illustrative example of this phenomenon can be witnessed in the upland regions of East and Southern Africa, where the escalating population pressure has acted as a catalyst for the severe degradation and desertification of areas that once held immense significance as crucial dry season grazing grounds for a multitude of pastoralist communities. As the relentless process of degradation persists unabated, it invariably paves the way for escalating conflicts and diminishing function of the resource. Inevitably, a disheartening outcome emerges where these previously invaluable lands become utterly useless to anyone, thereby resulting in the irrevocable loss of the resource for the original and deeply affected communities (Magan 2020; Karamalla-Gaiballa; Rahimi et al. 2021; Ariom et al. 2022; Ingty 2021; Mairomi and Kimengsi 2021).

4.2.7 Climate Change Impacts

This particular section aims to provide a concise and comprehensive summary that encompasses the projected impacts of climate change specifically for the African continent. By focusing on the agriculture system, which plays a pivotal role in African livelihoods and sustenance, we can gain a deeper understanding of the potential consequences that may arise. It is crucial to note that the majority of these effects are expected to have an adverse nature, with estimations suggesting that roughly 70% of the land in semi-arid areas may become unusable by the 2080s due to climate change. This significant decline in land usability could have far-reaching implications for agricultural practices and food production in these regions.

Furthermore, it is important to recognize that agricultural productivity in certain African countries may experience a substantial decrease, potentially up to 50%. An alarming consequence accompanying this decline is the projected devastating plummet in crop net revenues, estimated to be as severe as 90%. These predictions reflect the magnitude of the challenges that Africa's agricultural sector is expected to face in the coming years as a direct result of climate change. It is vital for policymakers, researchers, and communities to address these issues proactively and implement strategies that can mitigate the detrimental effects on the agriculture system and safeguard the livelihoods of the African population.

To achieve such mitigation, it will be imperative to invest in climate-smart agricultural practices that prioritize sustainability, resilience, and adaptation. This includes promoting the use of improved seeds and crop varieties that are more resistant to drought, pests, and diseases. Additionally, implementing efficient irrigation methods, soil conservation techniques, and precision farming technologies can contribute to conserving water, mitigating soil erosion, and optimizing resource allocation, thereby enhancing overall agricultural productivity. Furthermore, fostering agroforestry systems, promoting reforestation initiatives, and integrating livestock and crop production can contribute to ecosystem restoration, biodiversity conservation, and carbon sequestration. These nature-based solutions not only address the impacts of climate change but also provide multiple benefits such as increased food security, enhanced rural livelihoods, and improved ecosystem services.

To assist farmers and agricultural communities in adapting to the changing climate, it will be crucial to provide them with access to climate information, early warning systems, and relevant training and capacity-building programs. This knowledge and support can empower farmers to make informed decisions, optimize their farming practices, and adopt climate-resilient strategies. Additionally, facilitating access to financial services, insurance schemes, and market opportunities can help farmers manage risks, recover from climate-related losses, and enhance their economic resilience.

In order to achieve the necessary scale and impact, it is essential to foster collaboration and partnerships among governments, international organizations, civil society, and the private sector. By working together, it is possible to mobilize resources, share knowledge and best practices, and develop innovative solutions that address the complex and interconnected challenges of climate change in the agricultural sector. This includes supporting research and development initiatives, promoting technology transfer, and scaling up successful interventions across different regions and contexts.

In conclusion, climate change poses significant threats to agriculture in Africa, with potential adverse impacts on land usability, agricultural productivity, and food security. However, through proactive measures and targeted interventions, it is possible to build climate-resilient agricultural systems that not only mitigate the negative effects but also contribute to sustainable development goals and the well-being of African communities. The time to act is now, as the future of Africa's agriculture depends on our collective efforts to combat climate change and ensure a prosperous and resilient future for all (Osaliya 2021; Mensah et al. 2024; Shaban et al. 2023; Afzal and Ragab 2020; Srivastava et al. 2022).

Rural communities, which are often located in areas with limited access to healthcare resources, are disproportionately affected by the consequences of climate change. In particular, they face increased vulnerability to diseases that are exacerbated by shifts in temperature and rainfall patterns. Among these diseases, malaria stands out as the leading cause of both mortality and morbidity in Africa. Historically, highland areas in Africa (as well as in other continents) situated above 1500m have been considered relatively safe from malaria transmission. However, due to the impact of global warming, this perception might drastically change. It is estimated that the number of highland areas in Africa suitable for malaria transmission could more than double, climbing from 20% to a staggering 60%. This projected increase in transmission risk poses a significant threat to the health and well-being of rural communities in these regions. It underscores the urgent need for targeted interventions, strategic planning, comprehensive healthcare initiatives, and community engagement to actively combat and mitigate the potential devastating impacts of climate change on vulnerable populations. The effectiveness of these efforts relies on the collaboration of various stakeholders, including governments, NGOs, healthcare professionals, and local communities. Together, by prioritizing the allocation of resources, implementing preventive measures, and strengthening healthcare systems in these rural areas, we can ensure that this impending health crisis is averted. Furthermore, it is essential to raise awareness among the general population about the link between climate change and health, as well as the specific risks faced by rural communities. Education campaigns, public outreach programs, and advocacy efforts can play a crucial role in fostering understanding, empathy, and urgency in addressing this critical issue. By taking immediate action, we have the opportunity to safeguard the health and well-being of rural populations, preserve ecosystems, and contribute to a more sustainable future for all (Singer 2020; Kamaraj et al. 2023; Mertens 2024; Gratien 2022; Colwel 2023; Vidal 2023).

The potential effects and ramifications brought about by the ongoing phenomenon of climate change are bound to have far-reaching and intensified impacts on the availability of crucial resources and the overall state of food security. Looking ahead to the year 2020, it is projected that an alarming number of individuals, ranging between 75 million to a staggering 250 million people, will find themselves exposed to heightened levels of water stress. It is worth noting that this rise in water stress will manifest itself as a direct consequence of the ever-dwindling accessibility and availability of this vital resource. As water scarcity steadily worsens, an unfortunate reality will begin to take hold—one characterized by an ominous surge in the frequency of devastating droughts, particularly in the regions of East Africa and the Sahel.

Needless to say, the distressing nature of this prediction is further exacerbated by the increasingly unpredictable nature of water conditions in these areas. With the combined impact of reduced water availability, decreased accessibility, and heightened variability, it becomes abundantly clear that the detrimental effects of these circumstances will be proverbially felt far and wide. This, in turn, will severely hamper the ability to sustainably and sufficiently produce the necessary amounts of food, and subsequently, plunge millions of individuals who heavily rely on agriculture as their primary means of sustenance into dire straits. It is of paramount importance, therefore, that proactive and decisive measures are promptly implemented to effectively counter and mitigate these looming challenges. By doing so, we can actively safeguard and ensure the ongoing well-being, as well as the very survival, of the most vulnerable populations adversely affected by the immense impact of climate change on their livelihoods (du 2023; Liu et al. 2022; Isaacman and Musemwa 2021; Tuyishimire et al. 2022; De Wrachien et al. 2021; Bhaga et al. 2020; Neupane et al. 2022; Meddi and Eslamian 2021).

4.2.8 Population Growth

Shown as rising from 220 million in 1950 to 800 million to date in the low case and over 2 billion in the high case, Africa's population will continue to grow dramatically in all but the most extreme policy scenarios. While the rate of natural increase will decline in the future, there remains considerable momentum in the age structure for further growth for several decades into the future. Because of its magnitude and momentum, its close association with rural dependency, and differentials in the length of the reproductive period as well as contraceptive practice, population growth plays a major role in shaping the scenarios of future land use and food production in Africa. Much of the growth is expected to occur in areas of high agro-climatic potential so that there will be strong pressure to intensify land use in order to avoid further encroachment on natural vegetation and conservation areas. Even more important is the prospect in several countries of continued migration to natural resource-rich but environmentally fragile areas such as watersheds and forests in search of land and hard. A critical consideration is the extent to which public health, economic development, and family planning programs can reduce the forced and delayed components of future fertility, thus facilitating migration to urban areas and reducing the total adverse impact of population growth on the environment. High mortality from AIDS would slow population growth but might increase inequalities in land distribution and impede the accumulation of natural and human capital in several severely affected countries.

As Africa's population continues to soar, expected to rise from 220 million in 1950 to a staggering 800 million in the present low case scenario and over an astounding 2 billion in the high case, it is evident that growth will persist in all but the most extreme policy scenarios. Although the rate of natural increase is projected to decline in the future, there exists significant momentum in the age structure, paving the way for continuous expansion for several decades to come. Given its magnitude and momentum, coupled with its close association with rural dependency and variations in the length of the reproductive period and contraceptive practices, population growth assumes a pivotal role in shaping the future of land use and food production in Africa. Anticipated growth is concentrated in regions boasting high agro-climatic potential, exerting immense pressure to maximize land utilization to prevent further encroachment upon precious natural vegetation and conservation areas. However, the most critical concern lies in the prospect of ongoing migration to natural resource-rich yet environmentally fragile territories like watersheds and forests, driven by the pursuit of land and resources. It is crucial to consider the degree to which public health initiatives, economic development efforts, and family planning programs can mitigate the involuntary and delayed aspects of future fertility, thus facilitating migration to urban areas and curtailing the overall adverse impact of population growth on the environment. While high mortality rates resulting from AIDS could potentially decelerate population growth, they may inadvertently exacerbate inequalities in land distribution and hinder the accumulation of both natural and human capital in severely affected countries (Ezeh et al. 2020; Vollset et al. 2020; Weber et al. 2020; Branca et al. 2021; Molotoks et al. 2021; Anyanwu and Salami 2021; Coates et al. 2021; Sanchez et al. 2021).

4.3 Linking Conservation and Food Security

The fundamental issue facing Africa today is how to achieve sustainable rural development. This is because most of the people residing on the continent, especially those who live in or near protected areas, are rural, and the economies of the majority of African countries heavily rely on sectors that are based on natural resources. As highlighted by Barrett and Arcese (1995), the efforts to promote rural development in Africa have primarily revolved around enhancing agricultural productivity. Frequently, the most effective means of dissuading communities from encroaching upon environmentally sensitive areas is to enhance their existing agricultural practices. However, there are proponents who argue that the most prudent way to enhance rural well-being while ensuring ecosystem integrity is to shift rural development strategies away from agriculture and towards the creation of alternative livelihood options. This concept has been framed within the context of the highly contentious land sparing versus land sharing scenario. It is widely accepted that by minimizing the space utilized by humanity, more room is available for the preservation of the non-human world (expressed in a multitude of scientific articles). Consequently, if conservation strategies were approached with more assertiveness in the context of rural development, the objective would be to foster smaller, yet financially stable and sustainable forms of development. Conservation organizations have attempted to demonstrate the economic benefits of maintaining intact ecosystem functionality as a means to achieve this. Needless to say, this can be an exceedingly challenging endeavor. To illustrate with a hypothetical example, convincing individuals engaged in rice cultivation within an area abundant in biodiversity to abandon their fields and embrace beekeeping represents a complex situation. While beekeeping represents a more environmentally friendly form of agriculture, a hive does not exclusively rely on a single plant species for foraging and will often have minimal requirements in terms of forested areas. It can be incredibly difficult to provide the same level of economic return as rice agriculture. As an alternative approach, revenue generated from beekeeping could be allocated to a fund that compensates households for collectively preserving the forest, with the hope that this may eventually become a viable pathway for development.

In conclusion, the challenge faced by Africa regarding sustainable rural development is of utmost importance. The continent's population is predominantly rural, and the economies of many African countries heavily depend on natural resources. Historically, efforts to promote rural development have centered around improving agricultural productivity. A proven approach has been to enhance existing agricultural practices, thereby discouraging communities from encroaching upon environmentally sensitive areas. However, there is an opposing viewpoint that suggests shifting rural development strategies away from agriculture and toward alternative livelihood options would better accomplish the goal of improving rural well-being while maintaining ecosystem integrity. This notion is often debated in the context of land sparing versus land sharing. Advocates for land sparing argue that minimizing the space utilized by humans allows for the preservation of non-human habitats. By adopting assertive conservation strategies within rural development, it is possible to foster smaller-scale yet financially stable and sustainable forms of development. Conservation organizations have endeavored to showcase the economic benefits of maintaining functional ecosystems to support this approach. Nevertheless, this poses significant challenges. To illustrate, consider the hypothetical scenario of convincing rice farmers in a biodiversity-rich area to transition to beekeeping. While beekeeping is more environmentally friendly, it does not rely on a single plant species for foraging and requires minimal forested areas. Providing an equivalent economic return to rice agriculture can be immensely challenging. As an alternative, revenue generated from beekeeping could be directed towards a fund that compensates households for collectively preserving the forest. Over time, this innovative approach could potentially become a viable pathway for rural development in Africa (Emeana et al. 2020; Yin et al. 2022; Abisuga-Oyekunle et al. 2020; Monau et al. 2020; Andreoni et al. 2021; Pawlak and Kołodziejczak 2020; Carmela et al. 2020; Viana et al. 2022; Adenle 2020).

4.3.1 Balancing Conservation and Agriculture

When contemplating the intricate and interconnected relationship between farmers and parks in the realm of conservation, it becomes abundantly clear that subsistence and small-scale farmers play a significant role in utilizing agricultural land and thereby exerting a profound influence on the surrounding environment. It is important to highlight that the majority of these farmers are primarily concerned with meeting their own sustenance needs, and any surplus production is often intertwined with the goal of accessing broader markets. However, the fate of these farmers, specifically whether they will be displaced due to forthcoming conservation initiatives, heavily depends on the agricultural potential of the specific area in question. In cases where the area showcases a high level of agricultural viability, it is highly probable that it will become the primary focus for wildlife organizations or conservation non-governmental organizations (NGOs) seeking to protect and preserve biodiversity (Campbell 1999).

The aftermath of this trajectory can result in the sudden expulsion of farmers from their ancestral lands, leading to the complete cessation of all agricultural activities and the subsequent conversion of the land into designated wildlife areas or conservancies. Unfortunately, these actions possess the potential to ignite far-reaching social and economic consequences for the affected farmers (Neumann 1998) and, in some dire cases, even incite vehement resistance against the eviction measures themselves. It is evident that such circumstances are far from ideal for all parties involved, thus necessitating the development and implementation of comprehensive strategies that effectively uphold both food security and sustainable agriculture within rural communities. This is of utmost importance to ensure that the mere consideration of displacing farmers residing in areas harboring regions of significant conservation value does not arise as a viable option.

The intricate and intertwined relationship between farmers and parks in the realm of conservation brings to light the crucial role played by subsistence and small-scale farmers in utilizing agricultural land and exerting a profound influence on the surrounding environment. It is worth noting that these farmers are primarily concerned with meeting their own sustenance needs, and any surplus production is closely linked to accessing broader markets. However, the future of these farmers, particularly whether they will be displaced due to upcoming conservation initiatives, rests heavily on the agricultural potential of the specific area in question. In cases where the area demonstrates high agricultural viability, it is highly likely to become the primary focus for wildlife organizations or conservation non-governmental organizations (NGOs) seeking to protect and preserve biodiversity (Campbell 1999).

The consequences of this trajectory can lead to sudden expulsion of farmers from their ancestral lands, resulting in the complete halt of all agricultural activities and the subsequent conversion of the land into designated wildlife areas or conservancies. Unfortunately, these actions can have far-reaching social and economic implications for the affected farmers (Neumann 1998; Hulme and Murdoch 2001) and may even provoke strong resistance against the eviction measures themselves in dire cases. It is clear that such circumstances are far from ideal for all parties involved, necessitating the development and implementation of comprehensive strategies that effectively uphold both food security and sustainable agriculture within rural communities. This is crucial to ensure that the mere consideration of displacing farmers residing in areas harboring regions of significant conservation value is not seen as a viable option (Kom et al. 2022a, b; Azadi et al. 2021; Bellon et al. 2020; Devaux et al. 2021).

4.3.2 Sustainable Land Use Practices

The approach utilized to investigate the decline of wildlife populations will also be implemented to thoroughly examine and analyze the transformations occurring in land use. It is an undeniable fact that only a handful of regions across the globe remain untouched by the advancements of human technologies and the influence of diverse cultures. Even in the most remote and secluded corners of our majestic planet, indigenous communities have successfully developed and nurtured distinctive methods of managing their surroundings with the primary objective of ensuring the sustainability of the vast array of non-domesticated biotic resources. These time-honored and traditional systems have long been esteemed as highly sustainable, reliant on a profound and intricate comprehension of the environment and its intricate web of living organisms. It is fascinating to witness the incorporation of cultural taboos, wisely implemented to protect and preserve specific species or regions from any harm or disruption. Nowhere is the sheer magnitude and significance of these systems more prominently displayed than on the breathtaking and varied continent of Africa, teeming with excellently diverse and often challenging environmental conditions. Consequently, the indigenous populations of Africa have consistently demonstrate their remarkable power to skillfully utilize the natural resources found within their surroundings with the utmost efficiency. However, it is with great regret that we must acknowledge the unfortunate reality that these time-honored and ageless systems, meticulously refined and perfected over countless generations by local communities, have all too often encountered severe disruptions and unwelcome alterations due to the influence of external factors. As a direct result of these disruptions, observable and alarming declines in wildlife populations and profound modifications within precious habitats have become readily apparent (Adade Williams et al. 2020; Dawson et al. 2021; Brondízio et al. 2021; Lam et al. 2020; Litvinenko et al. 2022; Patel et al.; Domínguez and Luoma 2020; Mazzocchi 2020).

Almost every country in Africa, a diverse continent rich in natural resources and cultural heritage, has experienced significant changes in land use in the recent past. These transformations have been driven by a myriad of forces, ranging from demographic shifts and economic factors to technological advancements and political decisions. Understanding the complexity of these processes is crucial in comprehending the far-reaching consequences they have had on Africa's biota, the diverse array of plant and animal species that inhabit its ecosystems.

However, unraveling the intricacies of these changes in land use is no easy task. They are multifaceted, intertwined with social, economic, and environmental dynamics that make it difficult to discern individual factors and their respective impacts. As a result, comprehensive studies and rigorous analyses are required to disentangle this complexity, shedding light on the intricate web of cause and effect.

To gain a comprehensive understanding of the implications of changing land use in Africa, it is imperative to compare and contrast different land use systems. By examining the varied approaches taken in different regions and assessing their respective impacts, researchers can uncover valuable insights into the consequences of land use change. Such a comparative analysis allows for the identification of successful strategies and can guide the development of sustainable land use scenarios tailored to specific biotic regions.

In many instances, predicting the future impacts of certain types of land use becomes necessary. By employing robust predictive models and conducting scenario analysis, researchers can estimate the potential outcomes of various land use practices. This proactive approach enables policymakers and stakeholders to make informed decisions and devise effective conservation strategies, mitigating potential negative impacts on the continent's biota.

Furthermore, the findings derived from wildlife decline research can be applied to understanding land use change. The interplay between declining wildlife populations and changing land use patterns provides valuable lessons for crafting sustainable development strategies. By integrating knowledge from these two domains, researchers can contribute to the formulation of policies and practices that strike a balance between human needs and the preservation of Africa's unique biodiversity.

Comprehending the complex processes and impacts of changing land use in Africa necessitates a multifaceted approach. Only through meticulous research, comparative analysis, prediction modeling, and the integration of findings from related fields can we truly grasp the consequences of these transformations. By doing so, we can foster the development of sustainable land use scenarios that preserve Africa's biota for future generations (Winkler et al. 2021; Herrmann et al. 2020; Kouassi et al. 2021a, b; Hong et al. 2021; Asabere et al. 2020; Molotoks et al. 2021; Thomas and Whittington 2023; Malek and Verburg 2020; Potapov et al. 2022; Dibaba et al. 2020).

4.3.3 Integrated Farming Systems

This section discusses a wide range of conservation strategies, thoroughly examining the numerous advantages and disadvantages associated with each individual strategy. It is critically important to always bear in mind the immense complexity of conservation and farming within African smallholder agriculture. The challenges faced by smallholder farmers are significant, and the potential for poverty to increase among them due to the misuse or absence of appropriate conservation strategies cannot be understated. These strategies play a crucial role in ensuring the sustainability of smallholder agriculture.

Considering the wide array of conservation strategies that smallholder farmers may employ, it is highly probable that the implementation of such strategies will lead to a reduction in the overall yield of the main cropping system. This reduction can occur through various means, including the allocation of land for set aside purposes, the reduction of external inputs, the adoption of organic farming practices, or the implementation of conservation cropping techniques. While these measures may result in short-term challenges, it is important to consider their long-term benefits. By preserving the natural resources and enhancing the ecosystem services, these strategies contribute to the resiliency and sustainability of smallholder agriculture in the face of various environmental and economic stresses.

Therefore, it is of paramount importance to conduct thorough research and actively develop alternative systems that not only possess strong environmental sustainability, but also effectively achieve food security for all. Efforts should be made to identify innovative approaches that promote the coexistence of conservation and farming practices. These alternative systems should provide smallholder farmers with the option to either sustain or even increase their income levels. This can be achieved through diversification of their agricultural activities, incorporating value-added processes, and accessing high-value markets for their products.

One such system that holds great promise is integrated farming. While integrated farming has been an established concept in many developed countries for quite some time, it is still relatively new in terms of research and development within the African context. Integrated farming can be defined in numerous ways, but for the sake of clarity, let us make use of the definition proposed by (Franz et al. 1999), who state: "Integrated farming system optimizes the benefits from biological, physical, and chemical interaction between the faunal and floral components". With this definition in mind, it is evident that integrated farming systems comprehensively capitalize on the incredible advantages that arise from the interaction between the faunal and floral components, allowing for a holistic approach to farming and ultimately yielding a plethora of benefits for all involved.

Integrated farming systems promote the integration of crop production, livestock rearing, and aquaculture in a harmonious and mutually beneficial manner. By combining different agricultural activities, farmers can effectively utilize available resources, minimize waste, and enhance nutrient cycling. For example, the integration of crop residues with livestock production can not only provide animals with feed but also contribute to the replenishment of soil nutrients. Moreover, the cultivation of certain crops with specific traits can attract beneficial insects, birds, or other fauna that assist in pest control, reducing the dependence on chemical pesticides.

In addition to these ecological benefits, integrated farming systems offer socio-economic advantages. They enable a diversified income stream for smallholder farmers, reducing their vulnerability to market fluctuations and climate-related uncertainties. The integration of livestock and crops can create synergy, where the by-products of one activity serve as inputs for another, thereby enhancing resource use efficiency and reducing production costs. Furthermore, integrated farming systems have the potential to create employment opportunities, support rural development, and contribute to poverty alleviation.

To ensure the successful implementation of integrated farming systems in the African context, it is necessary to provide technical support, capacity building, and access to appropriate technologies and market linkages for smallholder farmers. Collaborative efforts between farmers, researchers, extension services, policymakers, and other stakeholders are vital to promote knowledge sharing, facilitate innovation, and address the specific challenges faced by smallholder farmers in different regions. The establishment of supportive policies and institutional frameworks is also crucial to create an enabling environment for the adoption and scaling-up of integrated farming systems.

In conclusion, integrated farming systems hold great potential in addressing the complex challenges of conservation and farming within African smallholder agriculture. By harnessing the benefits of faunal and floral interactions, these systems offer a holistic approach to farming that enhances sustainability, resilience, and economic viability. However, further research, collaboration, and investments are required to unlock the full potential of integrated farming in improving the livelihoods of smallholder farmers and ensuring food security for all (Corbeels et al. 2020; Ariom et al. 2022; Brown et al. 2020; Mgolozeli et al. 2020; Madembo et al. 2020; Rodenburg et al. 2021; Ngoma et al. 2021; Nyagumbo et al. 2020).

4.4 Case Studies

The Sango Bay area is located in a fertile valley in southern Tanzania and is characterized by smallholder agriculture. Numerous studies have been previously conducted in this particular area with the aim of examining the dynamics of agricultural land use and the ecological impacts that result from it on the surrounding environment. The main objective of these studies has been to develop a comprehensive toolkit that could be effectively utilized by farmers in similar areas to make well-informed and sustainable land use decisions that are not only economically viable but also environmentally beneficial. In order to achieve this goal, the study employed a combination of participatory rural appraisal and advanced spatial modelling techniques to gain a holistic understanding of the current land use patterns and the inherent trade-offs between profitability and environmental impact.

The primary finding of this groundbreaking study was that the current farming practices prevalent in the Sango Bay area allow for considerable economic returns. However, it has become increasingly evident that these practices are heavily reliant on an escalating amount of labor and fertilizer inputs. Therefore, it is of utmost importance to acknowledge that if external inputs, such as fertilizers, become scarce or incur higher costs in the future, the overwhelmingly high dependency on these inputs may lead to the grim scenario of land abandonment and subsequent reversion to natural vegetation.

This alarming trend of potential land abandonment can be attributed to the simultaneous decline in soil fertility and the escalating infestation of pests and diseases. Hence, it is crucial for farmers, policymakers, and other stakeholders to take urgent and proactive measures to combat these pressing challenges posed by declining soil quality and the onslaught of pests and diseases. Implementing effective soil conservation and fertility management practices, such as enhanced organic matter incorporation and crop rotation, alongside integrated pest management strategies, will undoubtedly play a pivotal role in ensuring the long-term sustainability and resilience of agricultural systems in the Sango Bay area and beyond.

As it is evident from this study, the dynamic relationship between agricultural productivity, environmental sustainability, and socioeconomic factors is complex and multifaceted. Therefore, the development and dissemination of this toolkit can serve as an invaluable guide for farmers in similar regions, empowering them to make informed decisions that strike a balance between economic gains and environmental preservation. By wholeheartedly embracing sustainable agricultural practices and embracing the principles of agroecology, farmers can foster a harmonious coexistence between their agricultural activities and the surrounding ecosystem, thereby ensuring not only their own welfare but also the conservation of biodiversity and the protection of natural resources for future generations to come.

The future of the Sango Bay area depends on the collective efforts and dedication of all stakeholders involved. By incorporating the knowledge and insights gained from this study, farmers can adapt their practices accordingly and contribute to the overall resilience and sustainability of the region. Additionally, policymakers can utilize the findings to develop targeted policies and interventions that support farmers in implementing sustainable agricultural practices. Moreover, raising awareness among the general public about the importance of sustainable land use and its positive impacts on the environment can encourage consumer support for locally produced, environmentally friendly products.

In conclusion, the significance of this study extends far beyond the boundaries of the Sango Bay area. It serves as a testament to the power of interdisciplinary research and participatory approaches in addressing complex agricultural and environmental challenges. The development and dissemination of the comprehensive toolkit highlighted in this study can be a catalyst for positive change, not only in similar rural regions but also in global efforts towards sustainable agriculture and environmental conservation. By working together, we can create a future where agricultural productivity thrives, natural resources are protected, and the well-being of both present and future generations is safeguarded (Matovu et al. 2024; Castella et al. 2023; Matovu et al. 2023; Nyamweya et al. 2023; Mulatu et al. 2022; Adetunji and MacKee 2023; Nseka et al. 2022; Stephenson et al. 2020; Sango 2021; Birungi 2022).

In Zimbabwe, a substantial amount of research has been conducted on wildlife utilization on private land, shedding light on the immense potential of this practice. The entire country has recognized wildlife-based land use as an incredibly viable alternative to conventional agricultural practices, primarily due to its ability to generate significantly higher economic returns while causing minimal harm to the environment. As a result, there is a widespread belief that this approach can help transform large areas of previously unproductive agricultural land back into their natural and lush vegetation states. This transformation, in turn, would play a crucial role in conserving the country's diverse ecosystem and creating designated wildlife areas that serve as safe havens for numerous endangered species.

One particular study, focusing on the beautiful savannah region of Zimbabwe, has effectively demonstrated the immense value of wildlife-based land use in achieving both economic prosperity and conservation objectives. By utilizing an integrated model framework, the study conclusively revealed that a well-managed wildlife-based system can generate returns equivalent to those of traditional cattle production, if not surpass them. The inherent advantages of this system are primarily derived from the access it provides to lucrative high-value export markets, combined with a relatively low level of domestic competition. As a result, wildlife-based land use emerges as an immensely lucrative venture that not only matches the profitability of conventional agricultural practices but also surpasses them in terms of utility.

Furthermore, the study found that wildlife-based land use has only a marginal impact on the current and future states of the environment, highlighting its sustainability and long-term viability. This stands in stark contrast to intensive agricultural practices, which tend to impose a heavy toll on the environment, leading to rapid land degradation and the exhaustion of valuable natural resources. In comparison, wildlife-based land use presents a model that strikes a harmonious balance between human activities and environmental preservation, making it a more environmentally responsible and sustainable approach in the long run.

In conclusion, the extensive research undertaken in Zimbabwe unequivocally emphasizes the transformative potential of wildlife-based land use. Apart from being an economically rewarding endeavor, it serves as a powerful instrument in conserving Zimbabwe's valuable biodiversity while allowing previously unproductive agricultural land to flourish with natural vegetation. By embracing this model, Zimbabwe can create vast wildlife areas, serving not only as a vital refuge for endangered species but also as a beacon of hope for sustainable land management and environmental stewardship (Heermans et al. 2021; Holechek et al. 2020; Shoo et al. 2021; Kolawole et al. 2023a, b; Chardonnet et al. 2023; Mwakiwa 2022; Chausson et al. 2020; Hinsley et al. 2023; Kadigi et al. 2023).

4.5 Successful Conservation and Food Security Initiatives

It is an incredibly intriguing and thought-provoking question to ponder whether the achievement of unparalleled success in either the realm of food security or the vast domain of conservation initiatives should undeniably be assessed and evaluated based on their profound impact on the other entity. Moreover, it brings forth the concept of whether it is absolutely imperative to strategically implement and execute transformative endeavors that possess an inherent and indelible dual purpose. The remarkably remarkable African Wildlife Foundation, renowned for their unwavering commitment to conservation efforts, eloquently and unequivocally highlight and emphasize in their captivating introductory statement for an exceptionally insightful and monumental symposium dedicated entirely to this resounding and paramount subject matter, that the latter contingency is unequivocally paramount and of utmost importance. In no uncertain terms, they ardently and fervently emphasize and assert the criticality, urgency, and sheer necessity of comprehending and meticulously reconciling potential conflicts that may arise between the noble pursuit of conservation and the ever-pressing demands of development. Subsequently, they underscore and fervently advocate for the identification and implementation of astute and ingenious strategies, initiatives, and initiatives that are poised to generate mutually beneficial outcomes, in order for the esteemed community of conservationists, development practitioners, and esteemed policy makers to genuinely and unquestionably accomplish and achieve their noble and paramount goals. In essence, it is truly and undeniably indispensable, essential, and non-negotiable for conservation initiatives to wholeheartedly and genuinely benefit and make a positive difference to the lives and well-being of local communities, just as food security endeavors, meticulously and strategically targeted at sustainable resource utilization, must unequivocally have a resoundingly positive, remarkable, and everlasting impact on both the remarkable people and the awe-inspiring and irreplaceable forces of nature that surround us all. In this regard, it is of paramount importance to recognize that the robust and profound integration of sustainable agriculture practices, effective ecosystem management, and innovative conservation methodologies, alongside comprehensive training and community engagement initiatives, can undeniably contribute to cultivating a harmonious coexistence between nourishing and safeguarding not only our ever-expanding global population but also the countless magnificent and intricate networks of flora and fauna that collectively form the vibrant tapestry of our planet's natural heritage. Thus, by fostering a multifaceted and synergistic approach that galvanizes the power of scientific research, technological advancements, economic empowerment, and social inclusion, we can unquestionably pave the way towards a future in which both the pursuit of food security and the preservation of biodiversity flourish harmoniously, nurturing and uplifting humanity while safeguarding the delicate secrets and wonders of the natural world (Lindsey et al. 2020; Kashwan etal. 2021; Cockerill and Hagerman 2020; Tobias Ochieng et al. 2021; Damoah and Adu 2022; Benkenstein et al. 2020; Foyet and Mupeta-Muyanwa 2023a, b; Bennett et al. 2021; Mabele et al. 2022; Shiffman et al. 2021).

The most obviously successful conservation initiatives are those that make areas no-go zones for humans. This is usually achieved by designating land as national parks or some category of protected areas. National park status being proven to be the most effective form of protection. This has been widely criticized by western conservationists as "fortress conservation" whereby relatively wealthy organizations or governments preserve scenic wilderness areas by expelling resident poorer populations who have often been proven to be doing relatively low level damage to the resource. However, it is unjust to make a sweeping generalization as the creation of a park has in many cases increased the availability of food resources to local people. For example, the cessation of cattle grazing has allowed for regeneration of firewood, and wild meat has become more abundant as animals within the park lose their fear of humans. Additionally, the original objective of conservation has seen numerous successes, usually being characterized by at least slowed down rates of forest clearance and habitat loss, particularly in and around buffer zones of protected areas. These positive outcomes highlight the importance of effective conservation measures in mitigating the loss of biodiversity and preserving crucial ecosystems. By striking a delicate balance between preservation and the well-being of local communities, we can ensure a more sustainable future for both humans and the natural world. The harmonious coexistence of humans and their surrounding environment is pivotal for the survival of our planet. Therefore, it becomes imperative to develop comprehensive frameworks that incorporate ecological preservation alongside the sustainable development of local communities. A solid understanding of the intricate interplay between ecosystems, local economies, and cultural identities is crucial in crafting effective conservation strategies. This involves fostering a holistic approach that values the well-being of both humans and the diverse array of species that inhabit our planet. In doing so, we can forge a path towards a harmonious future where the integrity of ecosystems is conserved, and the livelihoods of local communities are safeguarded. It is through such endeavors that we can build a more resilient and thriving planet for generations to come.

The commitment to conservation efforts is vital for the sustainable future of our planet. It is widely acknowledged that successful initiatives involve establishing no-go zones for human activities, often by designating land as national parks or protected areas. National parks have proven to be the most effective form of protection, safeguarding fragile ecosystems and biodiversity. Nevertheless, some Western conservationists criticize this approach, labeling it as "fortress conservation" and accusing wealthier organizations or governments of displacing poorer communities who may only minimally harm the environment. However, it is important to avoid generalizations, recognizing that the creation of parks has often enhanced food resources for local populations. For instance, the cessation of cattle grazing has allowed for the regeneration of firewood, while increased abundance of wild meat has occurred as animals within the park become less fearful of humans.

Moreover, conservation initiatives have achieved significant successes by slowing down rates of deforestation and habitat loss, particularly in buffer zones surrounding protected areas. These positive outcomes underscore the critical role of effective conservation measures in mitigating biodiversity loss and preserving vital ecosystems. Striking a delicate balance between preservation and the well-being of local communities is essential to ensure a sustainable future for both humanity and the natural world. The harmonious coexistence of humans and their surrounding environment plays a pivotal role in the survival of our planet. To achieve this, comprehensive frameworks must be developed, integrating ecological preservation with the sustainable development of local communities.

Crafting effective conservation strategies requires a comprehensive understanding of the complex interplay between ecosystems, local economies, and cultural identities. Adopting a holistic approach that values the well-being of humans and the diverse array of species inhabiting our planet is crucial. Through such an approach, we can pave the way towards a harmonious future where ecosystem integrity is preserved, and the livelihoods of local communities are safeguarded. By undertaking these endeavors, we can build a more resilient and thriving planet for present and future generations alike. Let us remain committed to preserving the wonder and beauty of our natural world (Mekonen 2020; Spenceley et al. 2021; Esfandiar et al. 2020; Buxton et al. 2021; Zhang et al. 2020; Setiawan and Suriansyah 2023; Domínguez and Luoma 2020; Grantham et al. 2020; Buckley 2020; Dawson et al. 2021).

4.5.1 Lessons Learned from Failed Attempts

One of the lessons that can be drawn from these often 'top down' project discourses is that they generally tend to be applied universally across Africa, despite the huge differences even within countries, much less between them. This approach fails to take into account the many African success stories in agricultural and environmental resilience to be found in marginal cropping and pastoral areas, and instead imposes a set of ideals concerning cultivation and conservation which may be inappropriate to the local ecology and socio-political situation. A smallholder study from Northern Cameroon and Eastern Burkina Faso found that, in general, the farmers were quite open to conservation ideals, but they also recognized the complexity of their own circumstances. They acknowledged that the recent trend towards intensified land use was not driven solely by choice, but rather as a reaction to persisting tenure insecurity in areas of civil unrest, along with declining terms of trade for agricultural products. Heavily influenced by structural adjustment policies under the assumption that it will lead to export-led growth, macro-economic change in Africa during the 1980–1990s led to the removal of subsidies on agricultural inputs and a reduced investment in rural development funding. This significantly harmed smallholder agriculture and tried to force a transition to modern farming, despite the fact that customary systems were appearing far more successful in provisioning both food security and environmental health. Intensification efforts have sometimes led to what is known as "tragedies of the commons", where land considered to be "free" is open access for overuse and degradation because it is not individuals' primary land holding. Elinor Ostrom, a renowned scholar, is particularly vocal in her frustration at how poorly the commons is understood in many development projects. These projects have assumed that access to the commons can be utilized equally by all poor people, and they mistakenly believe that individual tenure is the main cause of degradation. Understanding the complexity and diversity within African countries is crucial when implementing agricultural and environmental initiatives. It is important to recognize and learn from the successful practices employed by African farmers in marginal cropping and pastoral areas. By doing so, we can avoid the imposition of ideals that may not be suitable for the local ecology and socio-political context. In Northern Cameroon and Eastern Burkina Faso, a study on smallholder farmers revealed their willingness to embrace conservation ideals. However, it was evident that the recent shift towards intensified land use was not a mere choice, but rather a response to ongoing tenure insecurity in regions affected by civil unrest, as well as the decline in terms of trade for agricultural goods. During the 1980–1990s, Africa experienced significant macro-economic changes driven by structural adjustment policies. These policies, aimed at fostering export-led growth, resulted in the removal of subsidies on agricultural inputs and reduced investment in rural development. Unfortunately, such measures had detrimental effects on smallholder agriculture, as they forced a transition to modern farming methods, disregarding the success of customary systems in ensuring food security and environmental sustainability. Additionally, the intensification efforts sometimes led to the occurrence of "tragedies of the commons," where land deemed as "free" was subjected to overuse and degradation due to its status as non-primary land ownership. Elinor Ostrom, a prominent voice in this field, vehemently expresses her frustration towards the lack of understanding regarding the commons in various development projects. These projects incorrectly assume that poor individuals can equally utilize common resources, disregarding the fact that individual tenure is not the sole cause of environmental degradation (Mittal; Diao and Takeshima 2020; Tinarwo 2021; Ndung'u 2023; Alaerts 2020; Johnston 2021; Mamogobo 2021; Dreiling 2020; Nafie 2021; Søndergaard 2023).

4.6 Policy Recommendations

A complex and multifaceted issue such as the dual challenge faced by Africa poses a considerable hurdle in finding viable solutions. Balancing the crucial aspects of environmental conservation and food security is often perceived as conflicting, as the former entails limiting agricultural usage of land. Even when policy recommendations are put forth, they frequently fall short in adequately addressing both food security and conservation goals. To tackle this predicament, one potential remedy is to adopt a strategy of land sparing, where the focus is on promoting high-intensity farming on already cleared land, thereby preserving additional land for conservation purposes. In theory, this approach aligns with practicality and holds promise for making significant strides in resolving both issues. However, there is a paucity of evidence to suggest that this is a realistic option for Africa. Given the existing population growth rates and their predicted trajectory in the future, it becomes evident that the demand for food in Africa will inevitably continue to surge, potentially outstripping the world market price for food. In such a scenario, it seems implausible that Africans would be willing to barter their land for inexpensive and possibly unreliable food imports. Consequently, there exists a genuine risk of witnessing a situation where agricultural intensification takes place on the most valuable remaining agricultural land, inevitably encroaching upon protected areas. For land sparing to achieve its intended objectives and successfully address the challenges of food security, meticulous planning is imperative. It demands a comprehensive strategy that goes beyond mere promotion of high-intensity farming, taking into account the intricate interplay between agricultural productivity, conservation efforts, and the overall sustainability of both food systems and ecosystems. Synergistic interventions that carefully consider the intricacies of the African context and engage stakeholders at various levels are crucial for ensuring the effectiveness of land sparing as a viable solution to the dual challenge faced by Africa. Furthermore, collaboration between governments, international organizations, and local communities is essential to overcoming the obstacles associated with implementing such a strategy. Furthermore, efforts must be made to incentivize farmers to adopt high-intensity farming techniques on already cleared land. This can be achieved through the provision of financial incentives, access to modern farming technologies, and the dissemination of knowledge and best practices. Additionally, capacity-building initiatives should be implemented to empower and educate farmers on sustainable farming methods, crop rotation, and water management techniques. By equipping farmers with the necessary resources and knowledge, they can maximize agricultural productivity without compromising environmental integrity. In addition to agricultural interventions, it is crucial to invest in alternative sources of food and energy. Research and development in the field of aquaculture, for instance, can provide an additional source of protein and alleviate the pressure on land resources. Similarly, investing in renewable energy infrastructure can reduce reliance on fossil fuels and mitigate the environmental impact of traditional energy sources. These diversification efforts can contribute to a more resilient and sustainable food system while reducing the strain on land resources. Moreover, fostering collaboration and knowledge-sharing platforms between different stakeholders is vital for the success of land sparing initiatives. Creating platforms for farmers, researchers, policymakers, and local communities to exchange ideas, experiences, and expertise can lead to innovative solutions tailored to the African context. Such collaboration can also drive the development of sustainable agricultural practices, the implementation of effective conservation strategies, and the promotion of inclusive governance frameworks. Ultimately, addressing the dual challenge of food security and environmental conservation in Africa requires a nuanced and context-specific approach. Land sparing, when implemented with careful consideration of social, economic, and environmental factors, has the potential to offer a viable solution. However, it must be accompanied by comprehensive planning, collaboration, and investment in alternative food and energy sources. By embracing a multifaceted approach and engaging all stakeholders, Africa can navigate the complexities of this dual challenge and pave the way for a sustainable and prosperous future (Loconto et al. 2020; Balmford 2021; Watteyn et al. 2020; Leakey 2020; Ruwanza et al. 2022; Giller et al. 2021; Lay et al. 2022; Visagie and Turok, 2020; Mizik, 2021; Atube et al. 2021).

4.6.1 Strengthening Collaboration

Given the limited resources that both conservation and agricultural interests have, it becomes even more imperative that they come together and find common ground to effectively address the pressing issues they mutually face. This seemingly simple yet crucial task involves a crucial recognition of the intricate interdependence between conservation and food security. By fostering a willingness among both conservation organizations and agricultural interests to pursue mutually beneficial solutions, they can strive towards a harmonious coexistence. Presently, the reality is far from ideal, as there exists a prevailing notion that agricultural interests and conservation organizations are adversaries. This detrimental perspective hinders progress and obstructs the possibility of finding sustainable solutions. Conversely, conservation organizations often claim victories over policies that seem to be detrimental to agricultural interests, further exacerbating the divide between the two sectors. This harmful state of affairs invariably affects both parties, impeding progress and jeopardizing the future of both sectors. In the long run, it is the conservation organizations that bear the brunt of these conflicts and face the repercussions. One distressing consequence of these conflicts is the encroachment of subsistence farmers into numerous national parks and reserves worldwide. Driven by political supporters and propelled by the lack of viable alternatives for the landless rural poor, these farmers have found a stronghold within these protected areas. Dislodging them without providing sustainable alternatives has proven to be an arduous task. Simultaneously, policies that have been implemented in an attempt to alleviate food security concerns have, at times, fallen short of their intended goals. Such policies, often predicated on ideological and urban biases rather than a comprehensive understanding of the needs and strategies of the rural poor, have failed to produce the desired outcomes. This glaring discrepancy indicates that both sectors have made errors in their decision-making processes by implementing policies that stem from attitudes rather than informed knowledge. The future of African conservation and food security hinges on the ability of both sectors to recognize and rectify these mistakes, and actively pursue sustainable solutions. This calls for a deeper understanding of the unique needs of Africa's rural majority. By shifting their focus from attitudes to knowledge-driven policies, conservation organizations and agricultural interests can forge a path towards genuine progress. Collaborative efforts must be made to address and overcome the challenges they face, while concurrently recognizing that the success of one sector is intricately intertwined with the success of the other. Only through a united front can these sectors navigate the obstacles that lie ahead and ensure a prosperous future for Africa's rich biodiversity, as well as the millions who depend on its resources for their livelihoods (Musakwa et al. 2020; Mayele and Woja 2022; Togoch 2020; Prosper 2021; Matanzima and Marowa 2022; Kachena and Shackleton 2024; Nkansah-Dwamena nd; Köpke et al. 2023a, b; Petruska 2023).

4.6.2 Investing in Agricultural Research

Agricultural research yields improved crop varieties, better crop management practices, and control of pests and diseases. These often have very high benefit-cost ratios, in some cases as high as 16:1. An analysis of 54 developing countries estimated an average return to investment in agricultural research of 32%. Rates of return were higher for research on food staples than for cash crops such as coffee and tea. Internal rates of return averaged about 50%; there were no significant differences between irrigated and rainfed areas. Many developing countries now spend less on running and maintaining their research systems than they did in the 1970s. The rate of growth of CGIAR funding slowed since then, and accelerated growth has resulted from donors' insistent pressure on the Consortium. National government investment slowed down in the developing countries, stagnated among the Eastern European and Central Asian economies in transition, and Eastern Asia also saw decreased investment. Sub-Saharan Africa, where more than one-third of the population still suffer from hunger and malnutrition, has experienced the most severe decline. Simulation studies for Sub-Saharan Africa have suggested that scaling back agricultural research by as little as 20% would increase food prices, and thus the number of malnourished children, by 11%. This would also release 75 million hectares of land (13% of which is forest, and 40% of which is fallow or pasture) from use in agriculture, and so increase deforestation and habitat destruction for the advancing frontiers of agriculture. In addition to these effects, the expansion of agricultural research also brings long-term benefits to the economy and environment. By continuously improving crop varieties, researchers not only enhance crop yields but also contribute to enhanced food security and increased income for farmers. Moreover, the adoption of better crop management practices leads to more sustainable and efficient agricultural systems, reducing the negative environmental impacts of farming activities. With the control of pests and diseases, farmers can protect their crops from devastating losses, ensuring stable production and income. The impact of agricultural research on investment returns is undeniable. The average return of 32% underscores the effectiveness of allocating resources to this sector. Higher returns for research on food staples demonstrate the crucial role of such studies in enhancing food production and addressing global hunger challenges. The fact that irrigated and rainfed areas show similar rates of return indicates the broad applicability of research findings across different agricultural landscapes. Notably, the shift in investment patterns among developing countries, Eastern Europe, Central Asia, and Eastern Asia has had repercussions on agricultural research. The decline in funding and investment in Sub-Saharan Africa is particularly concerning, given the region's already prevalent hunger and malnutrition issues. Simulation studies send a clear warning that reducing agricultural research budgets by even a small percentage can lead to significant consequences, including higher food prices, increased malnourishment, and the release of large tracts of land from agricultural use. This, in turn, fuels deforestation and habitat destruction, endangering natural ecosystems. In conclusion, the expansion of agricultural research plays a vital role in improving crop production, food security, and the overall well-being of societies. Its positive impact is reflected in benefit-cost ratios, investment returns, and the sustainability of agricultural systems. However, it is crucial for governments, institutions, and donors to recognize the importance of continued support and investment in agricultural research, especially in regions where hunger and malnutrition persist. By prioritizing these efforts, we can mitigate the adverse effects of limited research funding and ensure a sustainable and prosperous future for agriculture and the planet (Alston et al. 2020; Baldos et al. 2020; Ruzzante et al. 2021; Yuan et al. 2021; Pawlak and Kołodziejczak 2020; Yuan et al. 2022; Ortiz-Bobea et al. 2021; Zheng and Ma, 2023; Monzon et al. 2021; Brookes and Barfoot 2020).

4.6.3 Improving Infrastructure

The problem with trying to improve infrastructure is that those ventures are fundamentally place-based, yet conservation and food production alike are activities that are greatly affected by the temporal dynamics of the environment. For example, the construction of a road through a forest may be quite ecologically benign by logging standards, yet in reality, the road may increase settlement and deforestation rates for many decades to come, leading to significant long-term consequences. Typically, the reasons that development projects fail to address conservation stem from a lack of integration of conservation concerns into overall sectoral policies and plans, and a lack of assessment of the indirect impacts of the project on the environment. However, there are potential solutions to these problems that suggest that donor and national agencies concerned with development can effectively address conservation needs without implementing separate conservation projects. For this to happen, it is crucial that infrastructure improvements are considered within the framework of overall sectoral policies and plans, ensuring that potential impacts are thoroughly assessed and taken into account. An understanding of the indirect impacts of infrastructure can be built into projects through the use of simulation models, allowing for a comprehensive evaluation of the potential consequences. To illustrate, let us take the example of a WWF study conducted for USAID on the Mahakam River Basin in Indonesian Borneo. In this study, a cellular automaton model was used to examine how road development would affect land use, and subsequently, how different land use patterns would have an impact on the river and the aquatic resources it sustains. By employing such sophisticated modeling techniques, researchers were able to gain valuable insights into the complex interactions between infrastructure development, land use, and environmental dynamics. The findings from this study served as a crucial guide for decision-makers, enabling them to make informed choices that prioritize both sustainable development and conservation goals. Therefore, the integration of conservation concerns into infrastructure development is not only essential but also feasible. By considering the indirect impacts and using simulation models to evaluate different scenarios, development projects can effectively incorporate conservation requirements. This approach ensures that infrastructure improvements go hand in hand with sustainable development, safeguarding the environment and its fragile ecosystems for generations to come. In conclusion, the importance of integrating conservation considerations into infrastructure development cannot be overstated. It is vital to recognize that the impact of such projects extends beyond the immediate construction phase. By acknowledging the long-term consequences and implementing measures to mitigate potential harm, development endeavors can harmoniously coexist with the natural environment. Through comprehensive assessment, simulation modeling, and informed decision-making, it is possible to strike a balance between progress and environmental preservation. This proactive approach will guide us towards a sustainable future where infrastructure developments actively support the health and longevity of our planet (Razmjoo et al. 2021; Hoban et al. 2021; Lv and Shang 2023; Petti et al. 2020; An et al. 2021; Sordello et al. 2022; Vishal et al. 2021; Hoban et al. 2022; Padder and Bashir 2023; Irawan and Hartoyo 2022).

4.6.4 Enhancing Market Access

Expanding access to markets for poor individuals is of utmost importance due to its potential to enhance economic opportunities and alleviate poverty. By providing avenues to trade one's labor for financial remuneration, urban, regional, and international markets serve as the epitome of financial emancipation. Consequently, an increased income not only guarantees financial stability but also reinforces food security, as it is widely recognized that relying on purchased food is a more dependable approach compared to subsisting solely on locally harvested produce. Remarkably, even during the peak of the hunger season, numerous impoverished individuals who depend on external food supplies are slightly benefitted by inflated staple prices resulting from production insufficiencies or localized famines. To facilitate the desired market accessibility, a plethora of feasible approaches can be implemented, encompassing the promotion of private sector growth, strategic public investment and intervention, as well as the introduction of innovative local institutional structures. These approaches include encouraging the establishment of micro-enterprises, such as small-scale shops or market stalls, that can cater to the specific needs of the local population. These businesses not only provide job opportunities but also ensure that essential goods and services are readily available within the community. Additionally, initiatives aimed at improving transportation infrastructure, such as building roads or expanding public transportation networks, can significantly enhance market connectivity. This would enable individuals from remote or marginalized areas to access larger markets, thereby expanding their potential customer base and increasing their chances of economic success. Furthermore, governments and international organizations can play a crucial role in enhancing market access for the poor. Public investment in agricultural research and development can lead to increased productivity and improved farming techniques, ultimately resulting in a surplus of agricultural products available for trade. This surplus can then be sold in regional or international markets, generating income for small-scale farmers and enhancing their overall economic well-being. Additionally, government intervention in the form of subsidies or price controls can help stabilize staple food prices, ensuring that even during times of production shortfalls or localized famines, the purchasing power of the poor is not severely affected. Innovative local institutional structures also have the potential to greatly enhance market access for the poor. For example, the establishment of community-based savings and credit groups can enable individuals to pool their resources and access loans for investment purposes. This allows them to start or expand their own businesses, ultimately increasing their income and economic opportunities. Moreover, the creation of farmer cooperatives or collective marketing groups can provide small-scale farmers with a unified voice and stronger bargaining power in negotiations with buyers, ensuring fair prices and better market access. In conclusion, expanding access to markets for poor individuals is a crucial step towards economic empowerment and poverty alleviation. By implementing various strategies such as private sector development, public investment and intervention, and innovative local institutional structures, the potential for increased market accessibility becomes achievable. This, in turn, leads to improved financial stability, reinforced food security, and enhanced economic opportunities for the poor (Wudil et al. 2022; Osabohien et al. 2022; Bozsik et al. 2022; Ningi et al. 2022; Nchanji and Lutomia 2021; March and Failler 2022; Gebre et al. 2021; Chanza and Musakwa 2022; Dabone et al. 2021; Ajefu and Abiona 2020).

Public investment in roads, transport, and market facilities is generally a pro-poor way of increasing market access and improving overall economic opportunities for individuals living in poverty. Extensive research conducted in Uganda has revealed that individuals from marginalized backgrounds are the most enthusiastic users of newly developed road networks. These individuals have been found to benefit significantly from reduced transportation costs, allowing them to allocate their savings towards essential activities and investment in additional food production, as well as other income-generating enterprises. In order to ensure that the benefits of road construction are maximized for the most vulnerable, it is crucial to involve poor individuals in the actual building process. This approach, previously referred to as 'food-for-work', has proven to be highly effective in channeling both money and food resources directly to those who are most in need. However, it is essential to establish careful mechanisms and arrangements that prevent any form of exploitation during this process to safeguard the welfare of the laborers. Enhancing accessibility to markets plays a vital role in encouraging private sector investment in various sectors such as input supply, food marketing, rural nonfarm employment, and diverse livelihood strategies. This, in turn, creates a sustainable pathway for poverty reduction and inclusive economic growth. Nonetheless, it is important to recognize that governmental bodies at both national and local levels possess significant regulatory powers in the realm of markets and trade. In certain instances, these regulations may unintentionally hinder the progress of poor individuals. Thus, continuous monitoring of public policies related to these areas is imperative to carefully assess and mitigate any potential negative impacts on the food security of marginalized populations. Furthermore, it is crucial to acknowledge the role of technology in revolutionizing market access for individuals living in poverty. The advent of mobile phones and internet connectivity has opened up new avenues for economic participation and financial inclusion. Farmers, for instance, can now access real-time market information, connect with buyers, and negotiate better prices for their produce. This increased connectivity has not only empowered individuals but has also fostered the growth of emerging digital platforms that offer innovative solutions to bridge the gap between consumers and sellers. Moreover, it is essential to promote entrepreneurship and skill development among marginalized individuals to enhance their economic opportunities. By providing training programs and access to credit facilities, these individuals can acquire the necessary knowledge and resources to start their own businesses. Encouraging the establishment of cooperatives and community-based organizations can also foster inclusive growth by facilitating collective action, sharing of resources, and leveraging economies of scale. In conclusion, public investment in roads, transport, and market facilities is a powerful tool for poverty reduction and inclusive economic growth. By involving marginalized individuals in the construction process and continuously monitoring public policies related to markets and trade, we can ensure that the benefits of these investments reach those who need them the most. Additionally, embracing technological advancements and promoting entrepreneurship among the vulnerable can further enhance their economic opportunities and contribute to their overall well-being (Ardizzi et al. 2020; Muldoon et al. 2023; Doherty 2021; Mutenyo et al. 2022; Morgado et al. 2022; Kisu 2022a, b; Omoding et al. 2020; Ssewanyana 2023; Saha et al. 2022).

Both conservation and food security are complex, interactive systems involving people and the environment. This essay has demonstrated that attention to environmental change is integral to the future: to the future of people's livelihoods, to the future of the environment itself, and to the future of food security. However, it's equally important to reiterate that the environments of the African continent are vital not just to global symbolic representations of wilderness and fauna, but also to many millions of people for whom a range of natural resources are all important elements in household or national food security strategies. Shifting cultivation, with its sustainable agricultural practices that prioritize both food production and environmental integrity, can have a profound impact on the preservation of biodiversity and the sustainable management of resources. This approach promotes a harmonious relationship between humans and nature, ensuring that future generations will also benefit from the gifts of the land. Alternatively, in the case of the Hukkle

Cultivators in Niger, they face the challenge of coping with erratic and insufficient rainfall. Despite these adversities, their unwavering dedication to ensuring a crop of millet each year has led to the expansion of cultivated land, albeit at the expense of the now degraded land. This highlights the need for innovative solutions that balance the immediate needs of food production with long-term sustainability. In Nigeria state, communities have devised creative strategies to address declining rangeland and limited cattle access to water. Through the production and sale of yogurt and careful crop herding, they have found ways to adapt and thrive in the face of environmental challenges. These poignant examples underscore the interconnectedness of environmental change, societal well-being, and future food security. To effectively navigate the intersection of conservation and development, it is crucial to comprehend the intricate and dynamic relationships between different forms of environmental/ecological change and the diverse food security options available to disparate societal groups. By adopting a comprehensive and inclusive approach that incorporates the wisdom of local communities, we can forge meaningful connections between conservation and development, ultimately securing a sustainable future for both people and the environment that sustains us.

In summary this chapter contributes to a comprehensive understanding of the intricate and multifaceted roles that conservation plays in ensuring food security. It has empirically demonstrated that, even in the absence of explicit food security initiatives, conservation organizations have been instrumental in safeguarding food security through the implementation of indirect strategies. Moreover, this research highlights that when confronted with major conservation challenges, particularly the alarming loss and degradation of natural habitats and the extinction of species, these organizations have adapted their strategies towards more interventionist approaches. However, in doing so, they face the crucial task of meticulously assessing the trade-offs involved, critically evaluating the potential adverse impacts on nature while carefully considering the beneficial contributions to human well-being. Consequently, this study underlines the urgent need for clear priority settings and meticulous planning to effectively reconcile conservation and food security objectives. Additionally, the author's findings reveal that the ongoing debate surrounding the detrimental effects of agriculture on conservation lacks a consensus among experts. Moreover, it elucidates that the most vulnerable groups affected by this complex interplay are the shifting cultivators and the communities forcibly displaced from natural protected areas. Addressing this issue represents an increasingly crucial, yet highly intricate challenge that necessitates innovative and sustainable solutions. The research further unveils that there exists a substantial potential for achieving win-win outcomes and synergies between conservation efforts and food security goals. Nonetheless, realizing these potential benefits requires skillful negotiation and a nuanced understanding of the intricate social and ecological mechanisms at play in each specific context, built upon a solid foundation of carefully established priority settings. In conclusion, this study provides compelling evidence that conservation initiatives are indispensable in ensuring food security (Du Plessis 2023; Gashure et al. 2022a, b; Ortega-Beltran and Bandyopadhyay 2021; Sujadmiko et al. 2023). By drawing attention to the interdependencies and complexities of these interconnected realms, it emphasizes the pressing need for informed decision-making, harmonized strategies, and collaborative efforts among diverse stakeholders. Ultimately, the successful integration of conservation and food security objectives will pave the way for a sustainable future, where nature thrives and human populations flourish in harmony. The holistic approach advocated by this research, which encompasses the need for long-term planning, capacity building, and community engagement, serves as a blueprint for achieving transformative change. Expanding knowledge in this field is crucial for adapting to the rapidly evolving dynamics of our global society. It is imperative that policymakers, researchers, and practitioners recognize the intrinsic value of conservation and its pivotal role in safeguarding both ecosystems and the well-being of present and future generations. Only through concerted interdisciplinary efforts can we navigate the myriad challenges ahead and forge a path towards a sustainable and resilient future. The time to act is now (Tables 4.1 and 4.2).

Table 4.1 Conservation Efforts in Africa

Country	Issue at stake	Conservation efforts made	Impact of the effort
Kenya	Poaching of elephants and rhinos	Strengthened anti-poaching laws, armed ranger patrols, wildlife corridors	Increased elephant and rhino populations, reduced poaching incidents
South Africa	Rhino poaching crisis	Dehorning rhinos, high-tech monitoring (drones, DNA tracking), anti-poaching units	Poaching incidents have declined, better security in Kruger National Park
Tanzania	Deforestation in the Eastern Arc Mountains	Reforestation programs, agroforestry initiatives, community-based forest management	Thousands of hectares of degraded land restored, improved biodiversity
Uganda	Mountain gorilla habitat destruction	Gorilla conservation through eco-tourism, regulated forest access	Gorilla populations have increased, tourism revenue supports conservation
Rwanda	Loss of biodiversity due to agriculture expansion	Large-scale reforestation, protected area expansion, sustainable land use policies	Increased forest cover, improved wildlife habitats
Gabon	Illegal logging and deforestation	Creation of 13 national parks, strict logging regulations, eco-tourism development	11% of the country's land protected, reduced illegal logging
Namibia	Human–wildlife conflict	Community-based conservation programs, sustainable tourism initiatives	Local communities benefit financially, reduced conflicts with wildlife
Democratic Republic of Congo	Poaching and habitat loss for bonobos and okapis	Strengthened ranger patrols, community conservation projects, anti-poaching units	Improved protection for endangered species, increased conservation awareness
Botswana	Elephant overpopulation and habitat destruction	Elephant translocation programs, expansion of protected areas	Reduced human–elephant conflicts, better habitat management
Madagascar	Loss of endemic species due to deforestation	Establishment of protected reserves, tree planting campaigns	Increased protection for endangered species like lemurs
Ethiopia	Land degradation and desertification	Large-scale afforestation projects, land restoration initiatives	Slowed desertification, improved soil quality
Morocco	Overfishing and marine ecosystem degradation	Creation of marine protected areas, enforcement of sustainable fishing policies	Fish stock recovery, healthier marine ecosystems
Zimbabwe	Illegal hunting and over-extraction of resources	Strict conservation laws, community-based wildlife management	Improved biodiversity, reduced poaching
Senegal	Mangrove deforestation and coastal erosion	Mangrove reforestation projects, integrated coastal management	Improved fish breeding grounds, reduced coastal erosion
Mauritius	Coral reef degradation due to climate change	Coral reef restoration projects, marine protected areas	Enhanced marine biodiversity, healthier coral reefs

Data Source: Compiled from World Wildlife Fund (WWF), International Union for Conservation of Nature (IUCN), African Wildlife Foundation (AWF) reports and publications, 2015–2022

Table 4.2 Population trends in selected African countries from 1950 to 2020

Country	1950	1960	1970	1980	1990	2000	2010	2020
Nigeria	37,860,000	45,138,000	55,982,000	73,423,000	95,617,000	122,283,000	159,424,000	206,140,000
Ethiopia	18,434,000	22,151,000	28,063,000	35,486,000	48,133,000	65,891,000	88,013,000	114,963,000
Egypt	21,834,000	27,798,000	33,306,000	38,793,000	52,442,000	67,348,000	83,005,000	102,334,000
Democratic Republic of the Congo	12,184,000	15,237,000	18,425,000	24,398,000	33,187,000	46,499,000	65,966,000	89,561,000
South Africa	13,683,000	17,063,000	22,045,000	28,049,000	36,794,000	44,448,000	50,133,000	59,308,000
Kenya	6,077,000	8,105,000	11,252,000	16,327,000	23,949,000	31,369,000	40,513,000	53,771,000
Ghana	5,036,000	6,635,000	8,559,000	11,683,000	15,421,000	19,912,000	24,965,000	31,073,000
Uganda	5,158,000	6,785,000	9,048,000	12,986,000	17,342,000	24,013,000	33,425,000	45,741,000
Algeria	8,823,000	11,923,000	14,917,000	18,670,000	25,715,000	30,199,000	35,423,000	43,851,000
Sudan	5,654,000	7,527,000	9,891,000	13,668,000	19,103,000	27,143,000	34,847,000	44,909,000
Angola	4,148,000	5,632,000	6,988,000	8,922,000	12,738,000	17,858,000	23,609,000	32,866,000
Tanzania	7,713,000	10,003,000	13,302,000	18,226,000	25,846,000	34,300,000	47,656,000	59,734,000
Morocco	8,986,000	11,626,000	15,635,000	19,300,000	24,987,000	28,710,000	32,381,000	36,911,000
Mozambique	6,442,000	7,926,000	9,548,000	12,120,000	14,938,000	18,364,000	23,929,000	31,255,000
Madagascar	4,084,000	5,051,000	6,734,000	9,038,000	12,209,000	15,926,000	21,315,000	27,691,000
Cameroon	4,466,000	5,702,000	7,684,000	10,901,000	14,769,000	17,872,000	21,513,000	26,545,000
Côte d'Ivoire	2,630,000	3,607,000	5,044,000	7,452,000	11,197,000	16,013,000	20,617,000	26,378,000
Niger	2,564,000	3,278,000	4,515,000	6,417,000	8,993,000	12,091,000	17,138,000	24,207,000
Burkina Faso	4,284,000	5,067,000	6,371,000	7,976,000	10,264,000	12,803,000	16,442,000	20,903,000
Mali	4,638,000	5,406,000	6,733,000	8,836,000	11,441,000	13,819,000	17,085,000	20,250,000
Malawi	2,881,000	3,680,000	4,963,000	6,947,000	9,790,000	12,252,000	15,904,000	19,129,000

Data Source: United Nations, Department of Economic and Social Affairs, Population Division (2022). World Population Prospects 2022: Summary of Results. New York: United Nations. Available at: https://population.un.org/wpp/

References

Abisuga-Oyekunle OA, Patra SK, Muchie M (2020) SMEs in sustainable development: their role in poverty reduction and employment generation in sub-Saharan Africa. Afr J Sci Technol Innov Dev 12(4):405–419. https://doi.org/10.1080/20421338.2019.1656428

Abubakari M, Ibrahim AS, Dosu B, Mahama M (2023) Sustaining the urban commons in Ghana through decentralized planning. Heliyon 9(6):e17192. https://doi.org/10.1016/j.heliyon.2023.e17192

Adade Williams P, Sikutshwa L, Shackleton S (2020) Acknowledging indigenous and local knowledge to facilitate collaboration in landscape approaches: lessons from a systematic review. Land 9(9):331. https://doi.org/10.3390/land9090331

Adam JN, Adams T, Gerber JD, Haller T (2021) Decentralization for increased sustainability in natural resource management? Two cautionary cases from Ghana. Sustainability 13(4):2181. https://doi.org/10.3390/su13042181

Adenle AA (2020) Assessment of solar energy technologies in Africa: opportunities and challenges in meeting the 2030 Agenda and Sustainable Development Goals. Energy Policy 137:111180. https://doi.org/10.1016/j.enpol.2019.111180

Aderibigbe OR, Ezekiel OO, Owolade SO, Korese JK, Sturm B, Hensel O (2022) Exploring the potentials of underutilized grain amaranth (*Amaranthus* spp.) along the value chain for food and nutrition security: a review. Crit Rev food Sci Nutr 62(3):656–669. https://doi.org/10.1080/10408398.2020.1825320

Adetunji OS, MacKee J (2023) Frameworks for climate risk management in cultural heritage: a systematic review of the state of the art. J Cult Herit Manag Sustain Dev 13(2):245–262. https://doi.org/10.1108/JCHMSD-10-2021-0185

Adeyanju S, O'Connor A, Addoah T, Bayala E, Djoudi H, Moombe K, Sunderland T (2021) Learning from community-based natural resource management in Ghana and Zambia: lessons for integrated landscape approaches. Int Forestry Rev 23(3):273–297. https://doi.org/10.1505/146554821833992810

Adeyonu A, Akerele D, Kehinde MO, Ologbon OAC, Akintayo O, Kolawole R (2024) Nexus between participation in non-farm enterprises and poverty among rural farm households: evidence from Nigeria. Int J Soc Econ 51(2):243–260. https://doi.org/10.1108/IJSE-04-2023-0284

African Wildlife Foundation (2020) AWF Conservation Impact Report. African Wildlife Foundation. https://www.awf.org

Afzal M, Ragab R (2020) Assessment of the potential impacts of climate change on the hydrology at catchment scale: modelling approach including prediction of future drought events using drought indices. Appl Water Sci 10. Article number 215. https://link.springer.com/article/10.1007/s13201-020-01293-1

Ajefu JB, Abiona O (2020) The mitigating impact of land tenure security on drought-induced food insecurity: evidence from rural Malawi. J Dev Stud 56(12):2169–2193. https://doi.org/10.1080/00220388.2020.1762862

Alaerts GJ (2020) Adaptive policy implementation: process and impact of Indonesia's national irrigation reform 1999–2018. World Dev sciencedirect.com

Alston JM, Pardey PG, Rao X (2020) The payoff to investing in CGIAR research. umn.edu

Amara E, Adhikari H, Heiskanen J, Siljander M, Munyao M, Omondi P, Pellikka P (2020) Aboveground biomass distribution in a multi-use savannah landscape in southeastern Kenya: impact of land use and fences. Land 9(10):381. mdpi.com

Amede T, Konde AA, Muhinda JJ, Bigirwa G (2023) Sustainable farming in practice: building resilient and profitable smallholder agricultural systems in sub–Saharan Africa. Sustainability. mdpi.com

Amelework AB, Bairu MW, Maema O, Venter SL, Laing M (2021) Adoption and promotion of resilient crops for climate risk mitigation and import substitution: a case analysis of cassava for South African agriculture. Front Sustain Food Syst 5:617783. frontiersin.org

Amirova E, Kuznetsov M, Khakimova E, Tolmacheva A (2020) Integrated development of digital agribusiness platform to support import substitution of food products. In: BIO web of conferences (Vol. 27). EDP Sciences, p 00055. bio-conferences.org

An H, Razzaq A, Nawaz A, Noman SM, Khan SAR (2021) Nexus between green logistic operations and triple bottom line: evidence from infrastructure-led Chinese outward foreign direct investment in Belt and Road host countries. Environ Sci Pollut Res 28(37):51022–51045

Andreoni A, Mondliwa P, Roberts S, Tregenna F (2021) Structural transformation in South Africa: the challenges of inclusive industrial development in a middle-income country. oapen.org

Anyanwu JC, Salami AO (2021) The impact of COVID-19 on African economies: an introduction. Afr Dev Rev. nih.gov

Ardizzi M, Wilson B, Hayhurst L, Otte J (2020) "People still believe a bicycle is for a poor person": features of "bicycles for development" organizations in Uganda and perspectives of practitioners. Sociol Sport J 38(1):36–49

Ariom TO, Dimon E, Nambeye E, Diouf NS, Adelusi OO, Boudalia S (2022) Climate-smart agriculture in African countries: a Review of strategies and impacts on smallholder farmers. Sustainability 14(18):11370. mdpi.com

Asabere SB, Acheampong RA, Ashiagbor G, Beckers SC, Keck M, Erasmi S et al (2020) Urbanization, land use transformation and spatio-environmental impacts: analyses of trends and implications in major metropolitan regions of Ghana. Land Use Policy 96:104707. academia.edu

Asani J (2020) The role of leadership in migration patterns in the Southern African development community. uj.ac.za

Assefa T, Jha M, Reyes M, Worqlul AW, Doro L, Tilahun S (2020) Conservation agriculture with drip irrigation: effects on soil quality and crop yield in sub-Saharan Africa. J Soil Water Conserv 75(2):209–217. tamu.edu

Atube F, Malinga GM, Nyeko M, Okello DM, Alarakol SP, Okello-Uma I (2021) Determinants of smallholder farmers' adaptation strategies to the effects of climate change: evidence from northern Uganda. Agric Food Secur 10(1):1–14

Azadi H, Moghaddam SM, Burkart S, Mahmoudi H, Van Passel S, Kurban A, Lopez-Carr D (2021) Rethinking resilient agriculture: from climate-smart agriculture to vulnerable-smart agriculture. J Clean Prod 319:128602. cgiar.org

Baldos ULC, Fuglie KO, Hertel TW (2020) The research cost of adapting agriculture to climate change: a global analysis to 2050. Agricultural Economics 51(2):205–220. https://doi.org/10.1111/agec.12551

Balehegn M, Ayantunde AA, Amole TA, Njarui DMG, Nkosi BD, Müller FL, Adesogan AT (2022) Forage conservation in sub-Saharan Africa: review of experiences, challenges, and opportunities. Agron J 114(1):75–99. https://doi.org/10.1002/agj2.20927

Balmford A (2021) Concentrating versus spreading our footprint: how to meet humanity's needs at least cost to nature. J Zool 313(4):255–266. https://doi.org/10.1111/jzo.12836

Bellon MR, Kotu BH, Azzarri C, Caracciolo F (2020) To diversify or not to diversify, that is the question: Pursuing agricultural development for smallholder farmers in marginal areas of Ghana. World Dev 125:104682. https://doi.org/10.1016/j.worlddev.2019.104682

Benkenstein A, Chevallier R, Kosciulek D, Lebea D, Worth K (2020) Youth climate advocacy in Africa. South African Institute of International Affairs (SAIIA). https://saiia.org.za

Bennett NJ, Katz L, Yadao-Evans W, Ahmadia GN, Atkinson S, Ban NC, Wilhelm A (2021) Advancing social equity in and through marine conservation. Front Mar Sci 8:711538. https://doi.org/10.3389/fmars.2021.711538

Bezu S (2021) Rural development: the significance of agriculture and the non-farm economy. In: Aryeetey E, Devarajan S, Kanbur R, Kasekende L (eds) Handbook of development policy. Oxford University Press, pp 179–198

Bhaga TD, Dube T, Shekede MD, Shoko C (2020) Impacts of climate variability and drought on surface water resources in sub-Saharan Africa using remote sensing: a review. Remote Sens 12(24):4182. https://doi.org/10.3390/rs12244182

Birungi KJ (2022) Impact of implementation of a community-based wetland management plan on the conservation of Lake Mulehe wetland in Kisoro District, Uganda (Master's thesis). Kyambogo University

Bozsik N, Cubillos JP, Stalbek B, Vasa L, Magda R (2022) Food security management in developing countries: influence of economic factors on food availability and access. PLOS ONE 17(6):e0268702. https://doi.org/10.1371/journal.pone.0268702

Branca G, Arslan A, Paolantonio A, Grewer U, Cattaneo A, Cavatassi R, Vetter S (2021) Assessing the economic and mitigation benefits of climate-smart agriculture and its implications for political economy: a case study in Southern Africa. J Clean Prod 285:125161. https://doi.org/10.1016/j.jclepro.2020.125161

Brondízio ES, Aumeeruddy-Thomas Y, Bates P, Carino J, Fernández-Llamazares Á, Ferrari MF, Shrestha UB (2021) Locally based, regionally manifested, and globally relevant: indigenous and local knowledge for nature. Annu Rev Environ Resour 46:481–509. https://doi.org/10.1146/annurev-environ-012220-012127

Brookes G, Barfoot P (2020) GM crop technology use 1996–2018: farm income and production impacts. GM Crops Food 11(4):242–261. https://doi.org/10.1080/21645698.2020.1773198

Brown B, Nuberg I, Llewellyn R (2020) From interest to implementation: exploring farmer progression of conservation agriculture in Eastern and Southern Africa. Environ Dev Sustain 22:3159–3177. https://doi.org/10.1007/s10668-019-00365-5

Bruns B (2021) Polycentric solutions for groundwater governance in sub-Saharan Africa: encouraging institutional artisanship. Water 13(4):480. https://doi.org/10.3390/w13040480

Buckley R (2020) Nature tourism and mental health: parks, happiness, and causation. J Sustain Tour 28(9):1409–1424. https://doi.org/10.1080/09669582.2020.1742725

Buxton RT, Pearson AL, Allou C, Fristrup K, Wittemyer G (2021) A synthesis of health benefits of natural sounds and their distribution in national parks. Proc Natl Acad Sci 118(14):e2013097118. https://doi.org/10.1073/pnas.2013097118

Carmela Annosi M, Brunetta F, Capo F, Heideveld L (2020) Digitalization in the agri-food industry: the relationship between technology and sustainable development. Manag Decis 58(8):1737–1757. https://doi.org/10.1108/MD-09-2019-1329

Casprini D, Oppio A, Torrieri F (2023) Usi civici: open evaluation issues in the Italian legal framework on civic use properties. Land 12(3):676. https://doi.org/10.3390/land12030676

Castella JC, Lu J, Friis C, Bruun TB, Cole R, Junquera V, Vagneron I et al (2023) Beyond the boom-bust cycle: an interdisciplinary framework for analysing crop booms. Glob Environ Change 80:102651. https://doi.org/10.1016/j.gloenvcha.2023.102651

Chanza N, Musakwa W (2022) Revitalizing indigenous ways of maintaining food security in a changing climate: review of the evidence base from Africa. Int J Clim Change Strateg Manag 14(3):252–271. https://doi.org/10.1108/IJCCSM-05-2021-0060

Chardonnet P, Taylor R, Crosmary W, Tadjo SP, Ligate FA, Baldus R, Cornélis D (2023) Buffalo hunting: from a commodity to a high-value game species. Trop Conserv Sci 16:1–18. https://doi.org/10.1177/19400829231161785

Chausson A, Turner B, Seddon D, Chabaneix N, Girardin CAJ, Kapos V, Seddon N (2020) Mapping the effectiveness of nature-based solutions for climate change adaptation. Glob Change Biol 26(11):6134–6155. https://doi.org/10.1111/gcb.15310

Chinseu EL, Dougill AJ, Stringer LC (2022) Strengthening conservation agriculture innovation systems in sub-Saharan Africa: lessons from a stakeholder analysis. Int J Agric Sustainability 20(1):17–30. https://doi.org/10.1080/14735903.2021.1915987

Cisneros-Araujo P, Ramirez-Lopez M, Juffe-Bignoli D, Fensholt R, Muro J, Mateo-Sánchez MC, Burgess ND (2021) Remote sensing of wildlife connectivity networks and priority locations for conservation in the Southern Agricultural Growth Corridor of Tanzania (SAGCOT). Remote Sens Ecol Conserv 7(3):430–444. https://doi.org/10.1002/rse2.198

Coates MM, Sliwa K, Watkins DA, Zühlke L, Perel P, Berteletti F, Bukhman G (2021) An investment case for the prevention and management of rheumatic heart disease in the African Union 2021–2030: a modelling study. Lancet Glob Health 9(7):e957–e966. https://doi.org/10.1016/S2214-109X(21)00155-7

Cockerill KA, Hagerman SM (2020) Historical insights for understanding the emergence of community-based conservation in Kenya: international agendas, colonial legacies, and contested priorities. Ecol Soc 25(2), Article 17. https://doi.org/10.5751/ES-11450-250217

Colwel C (2023) Stuff: humanity's Epic journey from naked ape to nonstop shopper. Penguin Random House, New York

Corbeels M, Naudin K, Whitbread AM, Kühne R, Letourmy P (2020) Limits of conservation agriculture to overcome low crop yields in sub-Saharan Africa. Nature Food 1(7):447–454. https://doi.org/10.1038/s43016-020-0114-4

Dabone C, Mbagwu I, Muray M, Ubangha L, Kohoun B, Etowa E, Etowa J (2021) Global food insecurity and African, Caribbean, and Black populations during the COVID-19 pandemic: a rapid review. J Racial Ethn Health Disparities. Advance online publication. https://doi.org/10.1007/s40615-021-01073-0

Damoah B, Adu EO (2022) Environmental education in South African schools: the role of civil society organizations. Res Soc Sci Technol 7(1):1–18. https://doi.org/10.46303/ressat.2022.1

Dawson NM, Coolsaet B, Sterling EJ, Loveridge R, Gross-Camp ND, Wongbusarakum S, Rosado-May FJ (2021) The role of Indigenous peoples and local communities in effective and equitable conservation. Ecol Soc 26(3). Article 19. https://doi.org/10.5751/ES-12625-260319

De Wrachien D, Schultz B, Goli MB (2021) Impacts of population growth and climate change on food production and irrigation and drainage needs: a world-wide view. Irrig Drain 70(1):14–26. https://doi.org/10.1002/ird.2534

Devaux A, Goffart JP, Kromann P, Andrade-Piedra J, Polar V, Hareau G (2021) The potato of the future: opportunities and challenges in sustainable agri-food systems. Potato Res 64(4):681–720. https://doi.org/10.1007/s11540-021-09507-2

Diao X, Takeshima H (2020) Agricultural mechanization in Ghana: alternative supply models for tractor hiring services. In: Takeshima H, Diao X (eds) An evolving paradigm of agricultural mechanization development: How much can Africa learn from Asia? International Food Policy Research Institute. Washington, pp 377–400. https://doi.org/10.2499/9780896293755

Dibaba WT, Demissie TA, Miegel K (2020) Drivers and implications of land use/land cover dynamics in Finchaa catchment, northwestern Ethiopia. Land 9(4):113. https://doi.org/10.3390/land9040113

Doherty J (2021) Waste worlds: inhabiting Kampala's infrastructures of disposability. Durham NC: Duke University Press. https://doi.org/10.1515/9781478013384

Domínguez L, Luoma C (2020) Decolonising conservation policy: How colonial land and conservation ideologies persist and perpetuate Indigenous injustices. Land 9(3):65. https://doi.org/10.3390/land9030065

Dreiling MC (2020) Trade globalization and its consequences. In: Galbraith JK, Blaas J (eds) Challenges of globalization and prospects for an inter-civilizational world order. Springer, Cham, pp 263–291. https://doi.org/10.1007/978-3-030-31165-3_11

Du Plessis (2023) Water resources from a global perspective. South Africa's Water predicament: freshwater's unceasing decline. Springer International Publishing, Cham, pp 1–25 [HTML]

Emeana EM, Trenchard L, Dehnen-Schmutz K (2020) The revolution of mobile phone-enabled services for agricultural development (m-Agri services) in Africa: the challenges for sustainability. Sustainability 12(2):485. https://doi.org/10.3390/su12020485

Esfandiar K, Dowling R, Pearce J, Goh E (2020) Personal norms and the adoption of pro-environmental binning behaviour in national parks: An integrated structural model approach. J Sustain Tour 28(1):10–32. https://doi.org/10.1080/09669582.2019.1663203

Ezeh A, Kissling F, Singer P (2020) Why sub-Saharan Africa might exceed its projected population size by 2100. Lancet 396(10258):1131–1133. https://doi.org/10.1016/S0140-6736(20)32020-1

Falayi M, Gambiza J, Schoon M (2021a) A scoping review of environmental governance challenges in southern Africa from 2010 to 2020. Environ Conserv 48(3):161–173. https://doi.org/10.1017/S0376892921000197

Falayi M, Gambiza J, Schoon M (2021b) A scoping review of environmental governance challenges in southern Africa from 2010 to 2020. Environ Conserv 48(4):253–264. https://doi.org/10.1017/S0376892921000306

Fastovich GG, Kapsargina SA (2020) Institute of import substitution in the agro-industrial complex of modern Russia (theoretical and legal approach). In: IOP conference series: earth and environmental science. IOP Publishing, 421(3):032035. https://doi.org/10.1088/1755-1315/421/3/032035

Foyet M, Mupeta-Muyanwa P (2023a) Human rights-based conversation: the integral role of human rights defenders in the conservation sector. J Environ Law & Policy 3(1):45–63

Foyet M, Mupeta-Muyanwa P (2023b) Human rights-based conversation: the integral role of human rights directors in the conservation sector. J Environ Law & Policy 3(2):45–62

Franz W, Baumann E, Müller H (1999) Integrated farming systems: Concepts, principles, and applications. J Sustain Agricul 14(2–3):25–44. https://doi.org/10.1300/J064v14n02_04

Gargallo E (2020a) Community conservation and land use in Namibia: visions, expectations and realities. J South Afr Stud 46(3):563–580. https://doi.org/10.1080/03057070.2020.1733897

Gargallo E (2020b) Community conservation and land use in Namibia: visions, expectations and realities. J South Afr Stud 46(4):745–762. https://doi.org/10.1080/03057070.2020.1775284

Gashure S, Wana D, Samimi C (2022a) Impacts of climate variability and climate-smart agricultural practices on crop production in UNESCO-designated cultural landscapes of Konso, Ethiopia. Theor Appl Climatol 147(3–4):1211–1226. https://doi.org/10.1007/s00704-021-03869-7

Gashure S, Wana D, Samimi C (2022b) Impacts of climate variability and climate-smart agricultural practices on crop production in UNESCO-designated cultural landscapes of Konso, Ethiopia. Theor Appl Climatol 149(1–2):505–520. https://doi.org/10.1007/s00704-022-04019-8

Gathuku GN (2022) Factors influencing Conservation of Rothschild's Giraffe in Ruma National Park and Mwea National Reserve, Kenya (Master's thesis). Kenyatta University, Nairobi, Kenya

Gebre GG, Mawia H, Makumbi D, Rahut DB (2021) The impact of adopting stress-tolerant maize on maize yield, maize income, and food security in Tanzania. Food Energy Secur 10(4):e313. https://doi.org/10.1002/fes3.313

Gibbon GEM (2021a) Understanding spatial priorities for conservation and restoration in Kenya (Doctoral thesis). University of Kent, Canterbury, United Kingdom

Gibbon GEM (2021b) Understanding spatial priorities for conservation and restoration in Kenya (Doctoral dissertation). University of Kent, Canterbury, UK

Giller KE, Delaune T, Silva JV, Descheemaeker K, van de Ven G, Schut AG, van Wijk MT, Hammond J, Hochman Z, Taulya G, Chikowo R, Narayanan S, Kishore A, Bresciani F, van Ittersum MK (2021) The future of farming: Who will produce our food? Food Security 13(5):1073–1099. https://doi.org/10.1007/s12571-021-01184-6

Grantham HS, Duncan A, Evans TD, Jones KR, Beyer HL, Schuster R et al (2020) Anthropogenic modification of forests means only 40% of remaining forests have high ecosystem integrity. Nat Commun 11(1):5978. https://doi.org/10.1038/s41467-020-19493-3

Gratien C (2022) The unsettled plain: an environmental history of the late Ottoman frontier. Stanford University Press

Gümplová P (2023a) Common ownership or global commons? Reassessing Risse's common ownership of the earth thesis in light of climate crisis. Curr Chall Environ Philos 5:101–118

Gümplová P (2023b) Common ownership or global commons? Reassessing Risse's common ownership of the earth thesis in light of climate crisis. In: Current challenges of environmental philosophy. Routledge, pp 93–112

Harmsen HW (2021) Effectiveness of patrolling protected areas against bushmeat poaching through snaring in the Soysambu Conservancy, Kenya (Master's thesis). University of Nairobi, Kenya

Heermans B, Van Rooyen J, Fynn R, Biggs D, Lewis M, McNutt J (2021) Husbandry and herding: a community-based approach to addressing illegal wildlife trade in northern Botswana. Front Conserv Sci 2:675493. https://doi.org/10.3389/fcosc.2021.675493

Herrmann SM, Brandt M, Rasmussen K, Fensholt R (2020) Accelerating land cover change in West Africa over four decades as population pressure increased. Commun Earth Environ 1(1):40. https://doi.org/10.1038/s43247-020-00053-y

Hinsley A, Challender DWS, Masters S, Macdonald DW, Milner-Gulland EJ, Fraser J, Wright J (2023) Early warning of trends in commercial wildlife trade through machine-learning analysis of patent filing. Conservation Biology 37(1):e13934. https://doi.org/10.1111/cobi.13934

Hoban S, Archer FI, Bertola LD, Bragg JG, Breed MF, Bruford MW et al (2022) Global genetic diversity status and trends: towards a suite of Essential Biodiversity Variables (EBVs) for genetic composition. Biol Rev 97(4):1511–1538. https://doi.org/10.1111/brv.12852

Hoban S, Bruford MW, Funk WC, Galbusera P, Griffith MP, Grueber CE et al (2021) Global commitments to conserving and monitoring genetic diversity are now necessary and feasible. Bioscience 71(9):964–976. https://doi.org/10.1093/biosci/biab054

Holechek JL, Geli HME, Cibils AF, Sawalhah MN (2020) Climate change, rangelands, and sustainability of ranching in the Western United States. Sustainability 12(12):4942. https://doi.org/10.3390/su12124942

Hong C, Burney JA, Pongratz J, Nabel JE, Mueller ND, Jackson RB, Davis SJ (2021) Global and regional drivers of land-use emissions in 1961–2017. Nature 589(7843):554–561. https://doi.org/10.1038/s41586-020-03138-y

Hossain A, Islam MT, Maitra S, Majumder D, Garai S, Mondal M et al (2021) Neglected and underutilized crop species: are they future smart crops in fighting poverty, hunger and malnutrition under changing climate? In: Neglected and underutilized crops: towards nutritional security and sustainability. Springer, Cham, pp 1–50. https://doi.org/10.1007/978-3-030-55187-4_1

Imathiu S (2021a) Neglected and underutilized cultivated crops with respect to indigenous African leafy vegetables for food and nutrition security. J Food Secur 9(3):104–112. https://doi.org/10.12691/jfs-9-3-4

Imathiu S (2021b) Neglected and underutilized cultivated crops with respect to indigenous African leafy vegetables for food and nutrition security. J Food Secur 9(3):104–113. https://doi.org/10.12691/jfs-9-3-3

Ingty T (2021) Pastoralism in the highest peaks: Role of the traditional grazing systems in maintaining biodiversity and ecosystem function in the alpine Himalaya. PLoS One 16(1):e0245221. https://doi.org/10.1371/journal.pone.0245221

Irawan NC, Hartoyo E (2022) Environmental management and stakeholder roles in sustainable tourism development: a feasibility study. IOP conference series: earth and environmental science 1108(1):012068. https://doi.org/10.1088/1755-1315/1108/1/012068

Irwin DA (2021) The rise and fall of import substitution. World Dev 139:105306. https://doi.org/10.1016/j.worlddev.2020.105306

Johnston LA (2021) "Getting old before getting rich" and Africa: of what relevance is China's economic demography transition? J Afr Dev 23(1):1–24

Joshi D, Monterroso I, Gallant B, Perera K, Peveri V (2021a) A gender–natural resources tango: Water, land, and forest research. In: Advancing gender equality through agricultural and environmental research. Springer, Cham, pp 182–220. https://doi.org/10.1007/978-3-030-65805-4_8

Joshi D, Monterroso I, Gallant B, Perera K, Peveri V (2021b) A gender–natural resources tango: Water, land, and forest research. In: Advancing gender equality through agricultural and environmental research: Past, present, and future. International Food Policy Research Institute (IFPRI), pp 182–220. https://doi.org/10.2499/9780896294155_09

Juma LO, Bakos IM, Khademi-Vidra A (2020a) Nature interpretation and visitor management objectives: a survey of tourist attitudes at Maasai Mara National Reserve, Kenya. Sustainability 12(3):1100. https://doi.org/10.3390/su12031100

Juma LO, Bakos IM, Khademi-Vidra A (2020b) Nature interpretation and visitor management objectives: a survey of tourist attitudes at Maasai Mara National Reserve, Kenya. Sustainability 12(6):2438. https://doi.org/10.3390/su12062438

Kachena L, Shackleton RT (2024) The impact of the invasive alien plant Vernonanthura polyanthes on conservation and livelihoods in the Chimanimani uplands of Zimbabwe. Biol Invasions 26:1321–1338. https://doi.org/10.1007/s10530-023-03162-6

Kadigi RM, Mgeni CP, Kangile JR, Aku AOA, Kimaro P (2023) Can a legal game meat trade in Tanzania lead to reduced poaching? Perceptions of stakeholders in the wildlife industry. J Nat Conserv 76:126502. https://doi.org/10.1016/j.jnc.2023.126502

Kalvelage L, Bollig M, Grawert E, Hulke C, Meyer M, Mkutu K et al (2021) Territorialising conservation: community-based approaches in Kenya and Namibia. Conserv Soc 19(4):282–293. https://doi.org/10.4103/cs.cs_20_81

Kamaraj C, Ragavendran C, Prem P, Naveen Kumar S, Ali A, Kazmi A et al (2023) Exploring the therapeutic potential of traditional antimalarial and antidengue plants: a mechanistic perspective. Can J Infect Dis Med Microbiolo 2023:8824139. https://doi.org/10.1155/2023/8824139

Kameri-Mbote P, Kamunge I, Yatich JK (2021) Kenya country report. In: African soil protection law. Nomos Verlagsgesellschaft, pp 177–294. https://doi.org/10.5771/9783748911180-177

Karamalla-Gaiballa N (n.d.). … Of climate change on the conflict in the Western Sudanese Region Of Darfur: is climate change the cause of the crisis? academia.edu

Kariuki RW, Munishi LK, Courtney-Mustaphi CJ, Capitani C, Shoemaker A, Lane PJ, Marchant R (2021) Integrating stakeholders' perspectives and spatial modelling to develop future land-use scenarios in northern Tanzania. PLOS ONE 16(2):e0245516. https://doi.org/10.1371/journal.pone.0245516

Kashwan P, Duffy R, Massé F, Asiyanbi AP, Marijnen E (2021) From racialized neocolonial global conservation to inclusive and regenerative conservation. Environment: Sci policy Sustain Dev 63(4):4–19. https://doi.org/10.1080/00139157.2021.1924574

Kassis G (2023) To what extent do institutional arrangements shape the excludability of resource systems? Lessons from French farms. J Inst Econ [HTML]

Kehinde AD, Ogundeji AA (2023) Distributive impacts of non-farm income on output and farm income of cassava farmers in Southwestern Nigeria. Scientific African 20:e01586. https://doi.org/10.1016/j.sciaf.2023.e01586

Kenjio JWK (2020) Decolonizing land tenure systems in sub Saharan Africa: the path to modern land policy reforms. J Land Manag Apprais 9(2):45–59

Kisu FM (2022a) The effectiveness of policy frameworks on traffic congestion (Doctoral dissertation, Makerere University, Kampala, Uganda)

Kisu FM (2022b) The effectiveness of policy frameworks on traffic congestion (Doctoral dissertation, Makerere University, Kampala, Uganda). Makerere University. https://makir.mak.ac.ug

Kohnert D (2024a) The impact of industrialized nation's CO_2 emissions on climate change in sub-Saharan Africa: case studies from South Africa, Nigeria and the DR Congo. SSOAR – Social Science Open Access Repository. https://doi.org/10.18452/27232

Kohnert D (2024b) The impact of industrialized nation's CO_2 emissions on climate change in sub-Saharan

Africa: case studies from South Africa, Nigeria and the DR Congo. SSOAR – Social Science Open Access Repository. https://doi.org/10.18452/26677

Kolawole OD, Hambira WL, Gondo R (2023a) Agrotourism as peripheral and ultra-peripheral community livelihoods diversification strategy: insights from the Okavango Delta, Botswana. J Arid Environ 210:104897. https://doi.org/10.1016/j.jaridenv.2023.104897

Kolawole OD, Hambira WL, Gondo R (2023b) Agrotourism as peripheral and ultra-peripheral community livelihoods diversification strategy: insights from the Okavango Delta, Botswana. J Arid Environ 210:104938. https://doi.org/10.1016/j.jaridenv.2022.104938

Kom Z, Nethengwe NS, Mpandeli NS, Chikoore H (2022a) Determinants of small-scale farmers' choice and adaptive strategies in response to climatic shocks in Vhembe District, South Africa. GeoJournal 87(6):5031–5050. https://doi.org/10.1007/s10708-021-10539-4

Kom Z, Nethengwe NS, Mpandeli NS, Chikoore H (2022b) Determinants of small-scale farmers' choice and adaptive strategies in response to climatic shocks in Vhembe District, South Africa. GeoJournal 87(6):4375–4392. https://doi.org/10.1007/s10708-021-10484-7

Köpke S, Withanachchi SS, Pathiranage R, Withanachchi CR, Gamage DU, Nissanka TS et al (2023a) Human-elephant conflict in the Sri Lankan dry zone: investigating social and geographical drivers through field-based methods. GeoJournal 88(5):5153–5172. https://doi.org/10.1007/s10708-022-10697-5

Köpke S, Withanachchi SS, Pathiranage R, Withanachchi CR, Gamage DU, Nissanka TS et al (2023b) Human-elephant conflict in the Sri Lankan dry zone: investigating social and geographical drivers through field-based methods. GeoJournal 88(5):5153–5172. https://doi.org/10.1007/s10708-022-10671-9

Kouassi JL, Gyau A, Diby L, Bene Y, Kouamé C (2021a) Assessing land use and land cover change and farmers' perceptions of deforestation and land degradation in South-West Côte d'Ivoire, West Africa. Land 10(5):429. https://doi.org/10.3390/land10050429

Kouassi JL, Gyau A, Diby L, Bene Y, Kouamé C (2021b) Assessing land use and land cover change and farmers' perceptions of deforestation and land degradation in South-West Côte d'Ivoire. Land 10(5):429. https://doi.org/10.3390/land10040429

Krauss JE (2021a) Decolonizing, conviviality and convivial conservation: towards a convivial SDG 15, life on land? J Polit Ecol 28(1):398–421. https://doi.org/10.2458/jpe.2464

Krauss JE (2021b) Decolonizing, conviviality and convivial conservation: towards a convivial SDG 15, life on land? J Polit Ecol 28(1):653–670. https://doi.org/10.2458/jpe.3033

Kuiper M, Cui HD (2021a) Using food loss reduction to reach food security and environmental objectives: a search for promising leverage points. Food Policy 98:101959. https://doi.org/10.1016/j.foodpol.2020.101959

Kuiper M, Cui HD (2021b) Using food loss reduction to reach food security and environmental objectives: identifying leverage points. Food Policy 98:101965. https://doi.org/10.1016/j.foodpol.2020.101965

Kumar S, Giri TK (2020a) Interconnectedness between spirituality and sustainable development goals: evidence from an Indian perspective. Int J Indian Cult Bus Manag 20(1):133–149. https://doi.org/10.1504/IJICBM.2020.105864

Kumar S, Giri TK (2020b) Interconnectedness between spirituality and sustainable development goals: evidence from an India. Int J Indian Cult Bus Manag 20(1):133–149. https://doi.org/10.1504/IJICBM.2020.105382

Kuntai ST (2020) Awareness and implementation of the critical factors for sustainable tourism: a case of Maasai Mara National Reserve, Kenya (Master's thesis). Strathmore University, Nairobi, Kenya

Kurbatova SM, Vlasov VV, Aisner LY (2020) Impact of risks and threats on the region's food supply in the context of import substitution. In: E3S web of conferences (Vol. 161, p. 01089). EDP Sciences. https://doi.org/10.1051/e3sconf/202016101089

Kuyah S, Sileshi GW, Nkurunziza L, Chirinda N, Ndayisaba PC, Dimobe K, Öborn I (2021a) Innovative agronomic practices for sustainable intensification in sub-Saharan Africa: a review. Agron Sustain Dev 41:1–21. https://doi.org/10.1007/s13593-021-00710-0

Kuyah S, Sileshi GW, Nkurunziza L, Chirinda N, Ndayisaba PC, Dimobe K, Öborn I (2021b) Innovative agronomic practices for sustainable intensification in sub-Saharan Africa: a review. Agron Sustain Dev 41:1–21. https://doi.org/10.1007/s13593-021-00706-4

Lam DPM, Hinz E, Lang DJ, Tengö M, von Wehrden H, Martín-López B (2020) Indigenous and local knowledge in sustainability transformations research: a literature review. Ecol Soc 25(1), Article 3. https://doi.org/10.5751/ES-11305-250103

Lawi MB, Adeniyi OR, Omolehin R (2022) Non-farm livelihood diversification and Sub-Saharan Africa's rural structural transformation: a review. Fudma J Sci 6(4):95–100

Lay J, Anseeuw W, Eckert S, Flachsbarth I, Kubitza C, Giger M (2022) Taking stock of the global land rush: few development benefits, many human and environmental risks. (Analytical Report III). Centre for Development and Environment (CDE), University of Bern, CIRAD; German Institute of Global and Area Studies (GIGA)

Leakey RRB (2020) A re-boot of tropical agriculture benefits food production, rural economies, health, social justice and the environment. Nat Food 1(5):260–265. https://doi.org/10.1038/s43016-020-0076-z

Lee M, Gambiza J (2022) The adoption of conservation agriculture by smallholder farmers in southern Africa: a scoping review of barriers and enablers. J Rural Stud 91:23–34. https://doi.org/10.1016/j.jrurstud.2022.02.012

Lindsey PA, Allan JR, Brehony P, Dickman A, Robson A, Begg C, Bhammar H, Blanken L, Breuer T, Fitzgerald

KH, Flyman M, Gandiwa E, Gibbons DW, Jones JP, Keane A, Manvell R, Packer C, Schep S, Sinclair ARE et al (2020) Conserving Africa's wildlife and wildlands through the COVID-19 crisis and beyond. Nat Ecol & evolution 4(10):1300–1310. https://doi.org/10.1038/s41559-020-1275-6

Litsardou D, Klabatsea R (2021). Land management strategies for sustainable (re-)development of public land assets in protected areas. Land 10(9):945. https://doi.org/10.3390/land10090945

Litvinenko V, Bowbrick I, Naumov I, Zaitseva Z (2022) Global guidelines and requirements for professional competencies of natural resource extraction engineers: implications for ESG principles and sustainable development goals. J Clean Prod 338:130530. https://doi.org/10.1016/j.jclepro.2022.130530

Liu X, Liu W, Tang Q, Liu B, Wada Y, Yang H (2022) Global agricultural water scarcity assessment incorporating blue and green water availability under future climate change. Earth's Future 10(6):e2021EF002567. https://doi.org/10.1029/2021EF002567

Loconto A, Desquilbet M, Moreau T, Couvet D, Dorin B (2020) The land sparing–land sharing controversy: tracing the politics of knowledge. Land Use policy 96:104627. https://doi.org/10.1016/j.landusepol.2020.104627

Lugo-Morin DR (2020) Indigenous communities and their food systems: a contribution to the current debate. J Ethnic Foods 7(1):6. https://doi.org/10.1186/s42779-020-00043-6

Lv Z, Shang W (2023) Impacts of intelligent transportation systems on energy conservation and emission reduction of transport systems: a comprehensive review. Green Technol Sustain 1(2):100015. https://doi.org/10.1016/j.grets.2023.100015

Mabele MB, Krauss JE, Kiwango W (2022) Going back to the roots: Ubuntu: and just conservation in Southern Africa. Conserv Soc 20(3):214–226. https://doi.org/10.4103/cs.cs_90_21

Madembo C, Mhlanga B, Thierfelder C (2020) Productivity or stability? Exploring maize-legume intercropping strategies for smallholder Conservation Agriculture farmers in Zimbabwe. Agric Syst 185:102921. https://doi.org/10.1016/j.agsy.2020.102921

Magan J (2020) Assessing the potential of holistic management in the rural village of Mceula in the Zulukama region, Eastern Cape (Master's thesis). Stellenbosch University, South Africa

Mairomi HW, Kimengsi JN (2021) Governance dynamics in rangelands: an evaluation of the applicability of Ostrom's principles in the Western Highlands of Cameroon. Scientific African 14:e01032. https://doi.org/10.1016/j.sciaf.2021.e01032

Malek Ž, Verburg PH (2020) Mapping global patterns of land use decision-making. Global Environ Change 65:102170. https://doi.org/10.1016/j.gloenvcha.2020.102170

Malemba E, Chiumya D, Anyanwu IE, Camkin J (2021). Enabling policy environment for water operator partnerships in the Southern African Development Community. World Water Policy 7(1):65–84. https://doi.org/10.1002/wwp2.12030

Mamogobo LN (2021) Local economic development and poverty alleviation in rural areas of South Africa: a case of Makhuduthamaga Local Municipality, Limpopo Province (Doctoral dissertation). University of Limpopo, South Africa

Manyani CRS (2020) From livestock to game farming: Farmers' understandings of land-use change, sustainable agriculture and biodiversity conservation in the Ubuntu Municipality, South Africa (Master's thesis). Stellenbosch University, South Africa

March A, Failler P (2022) Small-scale fisheries development in Africa: lessons learned and best practices for enhancing food security and livelihoods. Marine Policy 137:104946. https://doi.org/10.1016/j.marpol.2021.104946

Mashapa C, Gandiwa E, Muboko N, Mhuriro-Mashapa P (2021) Land use and land cover changes in a human-wildlife mediated landscape of Save Valley Conservancy, southeastern Zimbabwe. J Anim Plant Sci 31(2):514–527

Matanzima J, Marowa I (2022) Human–wildlife conflict and precarious livelihoods of the Tonga-speaking people of north-western Zimbabwe. In: Mutandwa M, Gukurume T (eds), Livelihoods of ethnic minorities in rural Zimbabwe. Springer, pp 107–122. https://doi.org/10.1007/978-3-031-05044-5_6

Matovu B, Sarfo I, Bbira Y, Yeboah E, Muhoozi Y (2023) Navigating through complexity: profiling the main threats to sustainable tropical wetlands management and implications for livelihoods and ecosystem services. Research Square. https://doi.org/10.21203/rs.3.rs-2485126/v1

Matovu B, Sarfo I, Bbira Y, Yeboah E, Muhoozi Y, Lukambagire I (2024) Navigating through complexity by profiling the main threats to sustainable tropical wetlands management and governance: a case study of Mityana District, Uganda. Discov Environ 2(1):9. https://doi.org/10.1007/s44207-023-00047-6

Mayele JM, Woja VS (2022) Socio-economic and environmental factors influencing human-wildlife conflict in Nimule National Park, South Sudan. Int J Innovative Sci Res Technol 7(4):465–482

Mazzocchi F (2020) A deeper meaning of sustainability: insights from indigenous knowledge. Anthropocene Rev 7(3):191–203. https://doi.org/10.1177/2053019620921101

Meddi M, Eslamian S (2021) Uncertainties in rainfall and water resources in Maghreb countries under climate change. In: Eslamian S, Eslamian F (eds) African handbook of climate change adaptation. Springer, pp. 1–23. https://doi.org/10.1007/978-3-030-45106-6_77

Mekonen S (2020) Coexistence between human and wildlife: Nature the nature, causes and mitigations of human–wildlife conflict around Bale Mountains National Park, southeast Ethiopia. BMC Ecol 20:51. https://doi.org/10.1186/s12898-020-00319-1

Mensah JK, Akpoti K, Antwi-Ofosu E, Kabo-bah AT, Siabi EK, Asare A et al (2024) Evaluating climate change scenarios in the White Volta Basin: a statistical bias-correction approach. Phys Chem Earth, Parts A/B/C, 103:103584. https://doi.org/10.1016/j.pce.2023.103584

Mertens JE (2024) A history of malaria and conflict. Parasitol Res 123:1–15. https://doi.org/10.1007/s00436-023-08056-1

Mgolozeli S, Nciizah AD, Wakindiki IIC, Mudau FN (2020) Innovative pro-smallholder farmers' permanent mulch for better soil quality and food security under conservation agriculture. Agronomy 10(9):1321. https://doi.org/10.3390/agronomy10091321

Michailidis A, Lazaridou D (2020) Non-farm employment: A key challenge to achieve zero hunger. In: Zero hunger. Springer, Cham, pp 137–154. https://doi.org/10.1007/978-3-030-42659-1_9

Mizik T (2021) Climate-smart agriculture on small-scale farms: a systematic literature review. Agronomy 11(6):1096. https://doi.org/10.3390/agronomy11061096

Molotoks A, Smith P, Dawson TP (2021) Impacts of land use, population, and climate change on global food security. Food Energy Secur 10(1):e261. https://doi.org/10.1002/fes3.261

Monau P, Raphaka K, Zvinorova-Chimboza P (2020) Sustainable utilization of indigenous goats in Southern Africa. Diversity 12(1):20. https://doi.org/10.3390/d12010020

Montt G, Luu T (2020) Does conservation agriculture change labour requirements? Evidence of sustainable intensification in sub-Saharan Africa. J Agric Econ 71(3):688–706. https://doi.org/10.1111/1477-9552.12375

Monzon JP, Slingerland MA, Rahutomo S, Agus F, Oberthür T, Andrade JF et al (2021) Fostering climate-smart intensification for oil palm. Nat Sustain 4(7):595–603. https://doi.org/10.1038/s41893-021-00700-8

Morgado Ramirez DZ, Nakandi B, Ssekitoleko R, Ackers L, Mwaka E, Kenney L et al (2022) The lived experience of people with upper limb absence living in Uganda: A qualitative study. Afr J Disability 11:890. https://doi.org/10.4102/ajod.v11i0.890

Mpofu E, Radinger-Peer V, Musakwa W, Penker M, Gugerell K (2023) Discourses on landscape governance and transfrontier conservation areas: converging, diverging and evolving narratives. Biodivers Conserv 32(14):4597–4626. https://doi.org/10.1007/s10531-023-02669-6

Mugiyo H, Chimonyo VGP, Sibanda M, Kunz R, Masemola CR, Modi AT, Mabhaudhi T (2021) Evaluation of land suitability methods with reference to neglected and underutilised crop species: A scoping review. Land 10(2):125. https://doi.org/10.3390/land10020125

Mugiyo H, Chimonyo VGP, Sibanda M, Kunz R, Nhamo L, Masemola CR et al (2021) Multi-criteria suitability analysis for neglected and underutilised crop species in South Africa. PLoS One 16(1):e0244734. https://doi.org/10.1371/journal.pone.0244734

Muhammad I, Rafii MY, Ramlee SI, Nazli MH, Harun AR, Oladosu Y et al (2020) Exploration of bambara groundnut (Vigna subterranea (L.) Verdc.), an underutilized crop, to aid global food security. Agronomy 10(6):766. https://doi.org/10.3390/agronomy10060766

Mulatu DW, Ahmed J, Semereab E, Arega T, Yohannes T, Akwany LO (2022) Stakeholders, institutional challenges and valuation of wetland ecosystem services in South Sudan. Environ Manag 69(4):666–683. https://doi.org/10.1007/s00267-021-01580-7

Muldoon J, Cant C, Graham M, Ustek Spilda F (2023) The poverty of ethical AI: impact sourcing and AI supply chains. AI Soc 38:1341–1356. https://doi.org/10.1007/s00146-022-01415-4

Musakwa W, Gumbo T, Paradza G, Mpofu E, Nyathi NA, Selamolela NB (2020) Partnerships and stakeholder participation in national park management: Gonarezhou National Park, Zimbabwe. Land 9(11):399. https://doi.org/10.3390/land9110399

Mutenyo J, Buyinza F, Ssenono V (2022) Digital divides or dividends? ca'Including basic services in Afris Digitalization Agenda: Evidence from Uganda. Information Technology for Development 28(4):773–792. https://doi.org/10.1080/02681102.2022.2043167

Mwakiwa E (2022) The socio-economic impacts of wildlife crop-raiding: reconciling conservation and agriculture. In: Protected areas and tourism in Southern Africa. Cham: Springer International Publishing, pp 145–162. https://doi.org/10.1007/978-3-030-91557-6_9

Nafie N (2021) The social effect of the IMF policies post arab-spring in the middle east

Naglis-Liepa K, Proškina L, Paula L, Kaufmane D (2021) Modelling the multiplier effect of a local food system. Rural Sustainability Research 45(340):38–46. https://doi.org/10.2478/plua-2021-0005

Nchanji EB, Lutomia CK (2021) Regional impact of COVID-19 on the production and food security of common bean smallholder farmers in sub-Saharan Africa: implication for SDG's. Global Food Secur 29:100524. https://doi.org/10.1016/j.gfs.2021.100524

Ndung'u GJ (2023) Shifts in the European Union's aid strategies towards Kenya (Master's thesis). Kenyatta University, Nairobi, Kenya

Neglo KAW, Gebrekidan T, Lyu K (2021a) Determinants of participation in non-farm activities and its effect on household income: an empirical study in Ethiopia. J Dev Agric Econ 13(1):72–92. https://doi.org/10.5897/JDAE2020.1218

Neglo KAW, Gebrekidan T, Lyu K (2021b) The role of agriculture and non-farm economy in addressing food insecurity in Ethiopia: a review. Sustainability 13(19):10995. https://doi.org/10.3390/su131910995

Neupane D, Adhikari P, Bhattarai D, Rana B, Ahmed Z, Sharma U, Adhikari D (2022) Does climate change affect the yield of the top three cereals and food security in the world? Earth 3(1):45–71. https://doi.org/10.3390/earth3010004

Ngoma H, Angelsen A, Jayne TS, Chapoto A (2021) Understanding adoption and impacts of conservation agriculture in eastern and southern Africa: a review. Front Agron 3:709875. https://doi.org/10.3389/fagro.2021.709875

Ngwu EC, Onah VC (2023) Between Samir Amin and Claude Ake: towards a roadmap for Africa's development. Politikon 50(1):45–62. https://doi.org/10.1080/02589346.2022.2136825

Ningi T, Taruvinga A, Zhou L, Ngarava S (2022) Factors that influence household food security in Hamburg and Melani, Eastern Cape, South Africa. Afr J Sci Technol Innov Dev 14(4):1050–1058. https://doi.org/10.1080/20421338.2021.1950213

Nkansah-Dwamena E (n.d.) Lessons learned from community engagement and participation in fostering coexistence and minimizing human-wildlife conflict in Ghana. Trees, Forests and People. Advance online publication. https://doi.org/10.1016/j.tfp.2023.100332

Nseka D, Opedes H, Mugagga F, Ayesiga P, Semakula H, Wasswa H, Ologe D (2022) Implications of land use and cover changes on upper River Rwizi macro-watershed health in southwestern Uganda. In: Water conservation – inevitable strategy. Makerere University Press, pp 1–22

Ntukey LT, Munishi LK, Kohi E, Treydte AC (2022) Land use/cover change reduces elephant habitat suitability in the Wami Mbiki–Saadani wildlife corridor, Tanzania. Land 11(2):202. https://doi.org/10.3390/land11020202

Ntumva ME (2022) Farmer-pastoralist conflict management approaches in sub-Saharan Africa: insights into their strengths and pitfalls. J Afr StudDev 14(3):43–55. https://doi.org/10.5897/JASD2021.0638

Nusratovich SK, Asidaxon A (2023) Influence of the volume of industrial production in Uzbekistan on the import trend. Qo'qon Universiteti Xabarnomasi 9:35–40

Nyagumbo I, Mupangwa W, Chipindu L, Rusinamhodzi L, Craufurd P (2020) A regional synthesis of seven year maize yield responses to conservation agriculture technologies in eastern and southern Africa. Agriculture, Ecosystems & Environment 295:106898. https://doi.org/10.1016/j.agee.2020.106898

Nyamweya C, Lawrence TJ, Ajode MZ, Smith S, Achieng AO, Barasa JE, Nkalubo, W (2023) Lake Victoria: Overview of research needs and the way forward. J Great Lakes Res 49(3):684–699. https://doi.org/10.1016/j.jglr.2023.03.006

Occhiali G (2023) What's the catch? A review of the fiscal treatments of fisheries in sub-Saharan Africa. J Environ Dev 32(4):387–412. https://doi.org/10.1177/10704965231174836

Ojiambo P (2022) Project management practices and implementation of wildlife conservation projects: a case of Kamungi Conservancy, Makueni County, Kenya (Master's thesis). University of Nairobi, Kenya

Okigbo RN, Anyaegbu CF (2021) Underutilized plants of Africa: Potentials for food security and sustainable development. J Biol Nat 9(2):45–58

Omoding J, Walters G, Andama E, Carvalho S, Colomer J, Cracco M et al (2020) Analysing and applying stakeholder perceptions to improve protected area governance in Ugandan conservation landscapes. Land 9(6):207. https://doi.org/10.3390/land9060207

Ortega-Beltran A, Bandyopadhyay R (2021) Contributions of integrated aflatoxin management strategies to achieve the sustainable development goals in African. Global Food Secur 28:100495. https://doi.org/10.1016/j.gfs.2020.100495

Ortiz-Bobea A, Ault TR, Carrillo CM, Chambers RG, Lobell DB (2021) Anthropogenic climate change has slowed global agricultural productivity growth. Nat Clim Change 11(4):306–312. https://doi.org/10.1038/s41558-021-01000-1

Osabohien R, Ashraf J, De Alwis T, Ufua DE, Osabuohien E, Odularu G et al (2022) Social protection and food security nexus in the Global South: empirical evidence from West Africa. Contemporary Soc Sci 17(2):129–142. https://doi.org/10.1080/21582041.2021.1996900

Osaliya R (2021) Impact of land use/cover and climate change on surface water resources in semi-arid Lokok and Lokere catchments, Uganda (Doctoral dissertation). University of Nairobi, Kenya

Padder FA, Bashir A (2023) Scarcity of water in the twenty-first century: problems and potential remedies. Medalion J Med Res Nurs Health Midwife Particip 4(1):1–5

Paliwal R, Adegboyega TT, Abberton M, Faloye B, Oyatomi O (2021) Potential of genomics for the improvement of underutilized legumes in sub-Saharan Africa. Legume Sci 3(3):e69. https://doi.org/10.1002/leg3.69

Patel SK, Sharma A, Singh GS (2021) Traditional agricultural practices in India: an approach for environmental sustainability and food security. Energy Policy 156:112419

Pawlak K, Kołodziejczak M (2020) The role of agriculture in ensuring food security in developing countries: considerations in the context of sustainable food production. Sustainability 12(13):5488. https://doi.org/10.3390/su12135488

Petruska M (2023) Conservation complexities in the Mount Elgon region, Uganda. Routledge

Petti L, Trillo C, Makore BN (2020) Cultural heritage and sustainable development targets: a possible harmonisation? Insights from the European perspective. Sustainability 12(3):926. https://doi.org/10.3390/su12030926

Popoola JO, Aworunse OS, Ojuederie OB, Adewale BD, Ajani OC, Oyatomi OA et al (2022) The exploitation of orphan legumes for food, income, and nutrition security in sub-Saharan Africa. Front Plant Sci 13:782140. https://doi.org/10.3389/fpls.2022.782140

Potapov P, Hansen MC, Pickens A, Hernandez-Serna A, Tyukavina A, Turubanova S et al (2022) The global 2000–2020 land cover and land use change dataset derived from the Landsat archive: first results. Front Remote Sens 3:856903. https://doi.org/10.3389/frsen.2022.856903

Prosper D (2021) Peasant's perception of wild animal's crop-raiding in villages surrounding Udzungwa Mountain National Park, Tanzania (Master's thesis). Open University of Tanzania

Qiang W, Liu A, Cheng S, Kastner T, Xie G (2020) Agricultural trade and virtual land use: the case of China's crop trade. Land Use Policy 90:104280. https://doi.org/10.1016/j.landusepol.2019.104280

Ragasa C, Andam KS, Asante SB, Amewu S (2020) Can local products compete against imports in West Africa? Supply- and demand-side perspectives on chicken, rice, and tilapia in Ghana. Global Food Secur 25:100349. https://doi.org/10.1016/j.gfs.2020.100349

Ramutsindela M, Mickler D (2020) Global goals and African development. In: Ramutsindela M, Mickler D (eds) Africa and the sustainable development goals. Springer, pp 1–22. https://doi.org/10.1007/978-3-030-14857-7_1

Rayamajhee V, Paniagua P (2021) The Ostroms and the contestable nature of goods: beyond taxonomies and toward institutional polycentricity. J Instit Econ 17(4):587–603. https://doi.org/10.1017/S1744137420000581

Razmjoo A, Nezhad MM, Kaigutha LG, Marzband M, Mirjalili S, Pazhoohesh M, Piras G (2021) Investigating smart city development based on green buildings, electrical vehicles and feasible indicators. Sustainability 13(14):7808. https://doi.org/10.3390/su13147808

Reaser JK, Tabor GM, Becker DJ, Muruthi P, Witt A, Woodley SJ, Plowright RK (2021) Land use-induced spillover: priority actions for protected and conserved area managers. PARKS 27(1):57–68. https://doi.org/10.2305/IUCN.CH.2021.PARKS-27-1JKR.en

Rodenburg J, Büchi L, Haggar J (2021) Adoption by adaptation: Moving from conservation agriculture to conservation practices. Int J Agric Sustain 19(5–6):437–455. https://doi.org/10.1080/14735903.2021.1910067

Ruwanza S, Thondhlana G, Falayi M (2022) Research progress and conceptual insights on drought impacts and responses among smallholder farmers in South Africa: a review. Land 11(5):702. https://doi.org/10.3390/land11050702

Ruzzante S, Labarta R, Bilton A (2021) Adoption of agricultural technology in the developing world: a meta-analysis of the empirical literature. World Dev 146:105599. https://doi.org/10.1016/j.worlddev.2021.105599

Saha M, Varghese D, Bartindale T, Thilsted SH, Ahmed SI, Olivier P (2022) Towards sustainable ICTD in Bangladesh: Understanding the program and policy landscape and its implications for CSCW and HCI. Proc ACM Human-Computer Interact 6(CSCW1): 1–31. https://doi.org/10.1145/3512925

Salustri A (2021) Social and solidarity economy and social and solidarity commons: towards the (re)discovery of an ethic of the common good? Ann Public and Coop Econ 92(1):27–50. https://doi.org/10.1111/apce.12298

Sanchez RG, Kougias I, Moner-Girona M, Fahl F, Jäger-Waldau A (2021) Assessment of floating solar photovoltaics potential in existing hydropower reservoirs in Africa. Renew Energy 169:687–699. https://doi.org/10.1016/j.renene.2021.01.041

Sango CO (2021) Factors affecting public housing development projects In Nairobi County. kcau.ac.ke

Séogo W (2023) Non-farm income and household welfare in rural Burkina Faso: a focus on the household land tenure situation. Int J Soc Econ 50(6):799–817. https://doi.org/10.1108/IJSE-09-2022-0627

Setiawan F, Murhaini ES (2023) Examining the psychological implications of the national park paradigm on inequality and poverty. J ReAttach Ther Dev Divers 6(9s2):122–130

Shaban M, Ghehsareh Ardestani E, Ebrahimi A, Borhani M (2023) Climate change impacts on optimal habitat of Stachys inflata medicinal plant in central Iran. Sci Rep 13(1):6580. https://doi.org/10.1038/s41598-023-33419-6

Shiffman DS, Macdonald CC, Wallace SS, Dulvy NK (2021) The role and value of science in shark conservation advocacy. Sci Rep. 11:2305 https://doi.org/10.1038/s41598-021-81877-3

Shoo RA, Mtui EK, Kimaro JM, Kinabo NR, Lendii GJ, Kideghesho JR (2021) Wildlife management areas in Tanzania: vulnerability and survival amidst COVID-19. PARKS, 27(Special Issue):97–108. https://doi.org/10.2305/IUCN.CH.2021.PARKS-27-SIRA.en

Shrestha J, Subedi S, Timsina KP, Chaudhary A, Kandel M, Tripathi S (2020) Conservation agriculture as an approach towards sustainable crop production: a review. Farming Manag 5(1):7–15

Singer M (2020) Ecosyndemics: global warming and the coming plagues of the twenty-first century. In: Plagues and Epidemics. Routledge

Slaev AD, Daskalova D (2020) Complex property rights and Coasean bargaining in natural resource management. J Environ Policy & Plan 22(4):486–500. https://doi.org/10.1080/1523908X.2020.1726733

Søndergaard N (2023) Brazilian agricultural diplomacy in the 21st century: a public–private partnership. [HTML]

Sordello R, Busson S, Cornuau JH, Deverchère P, Faure B, Guetté A et al (2022) A plea for a worldwide development of dark infrastructure for biodiversity–Practical examples and ways to go forward. Landsc Urban Plan 219:104332. sciencedirect.com

Soumare A, Diedhiou AG, Kane A (2022) Bambara groundnut: a neglected and underutilized climate-resilient crop with great potential to alleviate food insecurity in sub-Saharan Africa. J Crop Improv [HTML]

Spenceley A, McCool S, Newsome D, Báez A, Barborak JR, Blye CJ et al (2021) Tourism in protected and conserved areas amid the COVID-19 pandemic. Parks 27:103–118. murdoch.edu.au

Srivastava RK, Mequanint F, Chakraborty A, Panda RK, Halder D (2022) Augmentation of maize yield by strategic adaptation to cope with climate change for a future period in Eastern India. J Clean Prod 339:130599. [HTML]

Ssewanyana R (2023) Art and creativity for HIV/AIDS awareness, prevention, and empowerment of young people in Uganda. dur.ac.uk

Stephenson PJ, Ntiamoa-Baidu Y, Simaika JP (2020) The use of traditional and modern tools for monitoring wetlands biodiversity in Africa: challenges and opportunities. Front Environ Sci 8:61. frontiersin.org

Sujadmiko B, Narazua Y, Natamihardja R, Meutia IF (2023) Asean challenges toward Supranational Organization. Russian Law J 11(5S):238–248. cyberleninka.ru

Tamasiga P, Onyeaka H, Akinsemolu A, Bakwena M (2023) The inter-relationship between climate change, inequality, poverty and food security in Africa: a bibliometric review and content analysis approach. Sustainability 15(3):2146. https://doi.org/10.3390/su15032146

Thomas MF, Whittington GW (2023) Environment and land use in Africa. Routledge. https://doi.org/10.4324/9781003264937

Tian J, Han Y, Shen J, Zhu Y (2022) Sustainable development of agriculture with sustainable water management: an empirical investigation of "Five Water Cohabitation" in Zhejiang Province, China. Environ Monitor Assess 194(8):594. https://doi.org/10.1007/s10661-022-10239-8

Tinarwo J (2021) Effectiveness of the food and nutrition security policy in Masvingo Province, Zimbabwe (Master's thesis). University of Johannesburg

Tobias Ochieng N, Elizabeth KN, Nigel LW (2021) Measuring conservation attitudes of local communities towards the African elephant (Loxodonta africana) in the Mara ecosystem. PLOS ONE 16(3):e0248820. https://doi.org/10.1371/journal.pone.0248820

Togoch KH (2020) Analysis of human–wildlife conflicts on livelihood diversification of communities living adjacent to Tokamnarok National Reserve, baringo County, Kenya (Master's thesis). University of Nairobi

Turner WC, Périquet S, Goelst CE, Vera KB, Cameron EZ, Alexander KA et al (2022) Africa's drylands in a changing world: Challenges for wildlife conservation under climate and land-use changes in the Greater Etosha Landscape. Glob Ecol Conserv 38:e02221. https://doi.org/10.1016/j.gecco.2022.e02221

Tuyishimire A, Liu Y, Yin J, Kou L, Lin S, Lin J et al (2022) Drivers of the increasing water footprint in Africa: The food consumption perspective. Sci Total Environ 809:152196. https://doi.org/10.1016/j.scitotenv.2021.152196

Tyrrell P, du Toit JT, Macdonald DW (2020) Conservation beyond protected areas: Using vertebrate species ranges and biodiversity importance scores to inform policy for an East African country in transition. Conserv Sci Pract 2(1):e136. https://doi.org/10.1111/csp2.136

Ullah I, Kim DY (2021) Inclusive governance and biodiversity conservation: evidence from sub-Saharan Africa. Sustainability 13(10):5605. https://doi.org/10.3390/su13105605

United Nations, Department of Economic and Social Affairs, Population Division (2022) World Population Prospects 2022: Summary of Results. United Nations, New York. Available at: https://population.un.org/wpp/

United Nations, Department of Economic and Social Affairs, Population Division (2022) World population prospects 2022: Summary of Results. United Nations, New York. Available at: https://population.un.org/wpp/

Vanlauwe B, Dobermann A (2020) Sustainable intensification of agriculture in sub-Saharan Africa: first things first. Front Agric Sci Eng 7(4):376–382. https://doi.org/10.15302/J-FASE-2020356

Viana CM, Freire D, Abrantes P, Rocha J, Pereira P (2022) Agricultural land systems importance for supporting food security and sustainable development goals: a systematic review. Sci Total Environ 806:150718. https://doi.org/10.1016/j.scitotenv.2021.150718

Vidal J (2023) Fevered planet: how diseases emerge when we harm nature. Verso Books

Visagie J, Turok I (2020) Getting urban density to work in informal settlements in Africa. Environ Urban 32(2):569–590. https://doi.org/10.1177/0956247820929878

Vishal V, Chandra D, Singh U, Verma Y (2021) Understanding initial opportunities and key challenges for CCUS deployment in India at scale. Resour Conserv Recycl 175:105829. https://doi.org/10.1016/j.resconrec.2021.105829

Vollset SE, Goren E, Yuan CW, Cao J, Smith AE, Hsiao T et al (2020) Fertility, mortality, migration, and population scenarios for 195 countries and territories from 2017 to 2100: a forecasting analysis for the Global Burden of Disease Study. Lancet 396(10258):1285–1306. https://doi.org/10.1016/S0140-6736(20)30677-2

Waarts YR, Janssen V, Aryeetey R, Onduru D, Heriyanto D, Aprillya ST et al (2021) Multiple pathways towards achieving a living income for different types of smallholder tree-crop commodity farmers. Food Secur 13(6):1467–1496. https://doi.org/10.1007/s12571-021-01205-4

Watteyn C, Fremout T, Karremans AP, Huarcaya RP, Azofeifa Bolanos JB, Reubens B, Muys B (2020) Vanilla distribution modelling for conservation and sustainable cultivation in a joint land sparing/sharing concept. Ecosphere 11(3):e03056. https://doi.org/10.1002/ecs2.3056

Weber T, Bowyer P, Rechid D, Pfeifer S, Raffaele F, Remedio AR et al (2020) Analysis of compound climate extremes and exposed population in Africa under two different emission scenarios. Earth's Future 8(9):e2019EF001473. https://doi.org/10.1029/2019EF001473

Winkler K, Fuchs R, Rounsevell M, Herold M (2021) Global land use changes are four times greater than previously estimated. Nat Commun 12:2501. https://doi.org/10.1038/s41467-021-22702-2

Wolsink M (2020) Distributed energy systems as common goods: Socio-political acceptance of renewables in intelligent microgrids. Renew Sustain Energy Rev 127:109841. https://doi.org/10.1016/j.rser.2020.109841

Wu JJX (2022) Natural resource governance in Qing China: lineage institutions and the making of common property regimes. Cardozo Int Comp Laww Rev [HTML]

Wudil AH, Usman M, Rosak-Szyrocka J, Pilař L, Boye M (2022) Reversing years for global food security: a review of the food security situation in Sub-Saharan Africa (SSA). Int J Environ Res Public Health 19(22):14836. mdpi.com

Yin X, Chen J, Li J (2022) Rural innovation system: revitalize the countryside for a sustainable development. J Rural Stud [HTML]

Yuan S, Linquist BA, Wilson LT, Cassman KG, Stuart AM, Pede V et al (2021) Sustainable intensification for a larger global rice bowl. Nat Commun 12(1):7163. nature.com

Yuan S, Stuart AM, Laborte AG, Rattalino Edreira JI, Dobermann A, Kien LVN et al (2022) Southeast Asia must narrow down the yield gap to continue to be a major rice bowl. Nat Food 3(3):217–226. nature.com

Zhang Y, Xiao X, Cao R, Zheng C, Guo Y, Gong W, Wei Z (2020) How important is community participation to eco-environmental conservation in protected areas? From the perspective of predicting locals' pro-environmental behaviours. Sci Total Environ 739:139889. [HTML]

Zheng H, Ma W (2023) Smartphone-based information acquisition and wheat farm performance: insights from a doubly robust IPWRA estimator. Electronic Commer Res researchgate.net

5 Sustainable Farming Practices in Africa

5.1 Introduction, Definition and Importance of Sustainable Agriculture

Sustainable farming practices have to be adopted in Africa on a wider scale than they are at present. The sustainable farming practices have to be passed on from generation to generation if they are to be successful. "Sustainable agriculture calls for a renewed and refurbished understanding, to establish farming systems that are not harmful to the environment, and that in the long run can sustain the process." This quote basically defines what sustainable farming is. This method of farming is a process that was proven to preserve the quality and the resources of the ecosystem. It is slow and more labour intensive than modern methods, but the end result is less harmful to the environment, with a long run vision to sustain the process. This is a way of farming that, with the right teaching and practice, could be passed on through the generations. It also gives the opportunity for Africans, specifically in the rural areas, the sufficiency that they yearn. The Western society's complex and deeply rooted problems with unsustainable way of farming can easily be conveyed upon sustainable farming models. The developing world has a chance to learn from the western mistakes, various education can be provided and a change in thought can be provided earlier rather than later. This will pave the way for a better future for the rural African farmers.

In order to understand how to farm sustainably in Africa, several sustainable farming systems on the various crops and livestock need to be formulated and thoroughly tested. Then trial and error is to be done with the farmers in the rural areas for them to understand and convey the various methods. The key here is that the practices are appropriate, affordable and workable with minimal effect on the environment, which will increase in sustainability. Then a new generation of farmers, having been taught, can slowly replace the old ones. This is not easy and should be deemed a long-term goal. The oft-forgotten vast knowledge of traditional farming systems and the various ways it can be conveyed into modern sustainable farming needs to be university researched and documented so it is not lost. This requires purposeful persistence and dedication with the aim for the better future at the forefront of the mind. With the right methods, teaching and a bit of good luck, the world farming systems in Africa in the year 2100 can be changed for the better, providing sustainability and reducing the effects (Żmija et al. 2020; Lindsjö et al. 2021; Uebersax et al. 2023; Ibrahim et al. 2021; Akinola et al. 2020).

Sustainable farming is a method of farming that allows future generations the ability to farm. This system is considered to be the most effective way to increase food security in developing nations. For instance, in 1992, the World Food Summit stated that sustainable development was the most effective way to increase food security in developing nations. In

E. D. Wiafe, *Wildlife Conservation and Agriculture Development in Africa*, Sustainable Development Goals Series, https://doi.org/10.1007/978-3-032-09233-5_5

relation to 1992, the most recent food security reports from the Food and Agriculture Organization of the United Nations would suggest that development of sustainable systems is more urgent than ever. In these reports, the FAO is directing more attention to developing nations, where food security is in a state of emergency due in part to food price and income instability. It is noted that price volatility has steadily been on the increase. The recent food price crisis followed by the global economic crisis has only served to drive more people into hunger and poverty, further highlighting the need for increased food security measures. Price volatility has detrimental impacts on the poor and hungry as it can disrupt livelihoods and food consumption patterns. An increase in sustainable development and food security measures would be expected to decrease price and income instability through an increase in the quantity and quality of food supply. This is because sustainable agriculture and rural development have the potential to enhance food supply conditions, by fostering increases in productivity and access to food. Increased investment in agriculture is another potential outcome of increased food security measures, as agriculture has proven to be an effective way to reduce poverty in nations where food security is a pressing issue (Campi et al. 2021; Clapp et al. 2022; Pawlak and Kołodziejczak 2020; Viana et al.2022).

African agriculture and, more importantly, the welfare of millions of smallholder farmers, are caught up in a complex system of interconnected variables. Poverty and food insecurity, low agricultural productivity, natural resource degradation, and declining environmental resilience are key problems that appear to be locked in a self-perpetuating cycle. This is often referred to as the downward spiral of rural poverty. Smallholder farmers are a heterogeneous group but have little choice but to use the resources available to them to generate an income and progress their welfare. Too often, this involves the unsustainable intensification of agriculture into fragile lands with the sole concern of generating income. This incurs environmental costs, reduced resilience, and often leads to food insecurity and poverty due to the degraded state of the resources on which the farmers depend. This type of agriculture also has repercussions in the wider economy and environment. It is said that degradation constrains both demand and supply of products necessary for sustainable development, thus entrenching poverty. The major threats to Africa's main asset, its array of natural resources and biodiversity, are another issue of great concern. The unsustainable use of Africa's natural resources exists both internally and externally to the continent. Many farmers have little access to markets and, as a result, only produce enough for subsistence and are forced to exploit surrounding resources (Irawan 2023; Deka and Goswami 2021).

Sustainable agriculture is an approach that attempts to rectify this situation. The term sustainable agriculture is a broad phrase that is used to describe farming practices that are at least compatible with the maintenance of social, ecological, and biological systems while not depleting the earth's resources. This is often referred to as triple bottom line accounting, i.e. the capacity to generate increasing standards of living, economic wealth, and social capital while not degrading the environment and eroding the resource base. The crucial question for Africa is whether sustainable agriculture can lead to an increase in productivity and household welfare for the millions of rural Africans currently living in poverty (Harwood 2020; Barrios et al. 2020; Wezel et al. 2020).

Two thirds of the labour force in Africa are employed in agriculture, but since 1990s, the sector has been in crisis. The causes of the present agricultural crisis are not rooted in a lack of participation by Africa in international agricultural markets. Evidence suggests that African producers are quite successful in lowering their costs and competing effectively in both domestic and international markets. The fundamental cause of the crisis seems to lie in a lack of sustained and broad-based investment in agricultural productivity. This has led to a situation whereby there has been some development in certain cash crops and market sectors, but the majority of small farmers are confronted with a slow but progressive deterioration in their productive resources, particularly the natural resource base (Yeboah and Jayne 2020; Daum and Birner 2020).

5.2 Soil Conservation Techniques

Cover cropping is a method of growing crops that cover the soil. It is often used to protect soil from erosion and improve its quality. Usually, the crops are not harvested but are tilled back into the soil. The plants used for cover cropping are usually fast-growing crops with herbicidal properties. An example of a cover crop would be the sesbania plant, which is known to quickly cover soil and take up to 342 kg of nitrogen per hectare. This method is known to sustain the soil by maintaining its fertility and organic matter in the long term.

Terracing is an alignment of crops around the landscape in a series of small steps. It has traditionally been used in Asia to create rice paddies on steep hillsides. It can be adapted to many different types of crops and has the effect of reducing water and soil movement downslope by breaking the slope length. This controls erosion quite effectively and improves water conservation in the soil.

Crop rotation is a very useful way of conserving soil. It is a technique that involves changing crops every season. A very good example of this in Africa is the planting of legumes and maize. Legumes have nodules in their roots that contain bacteria called Rhizobia. This bacteria is able to convert atmospheric nitrogen into a form that is usable by plants. This has a positive effect on the maize crop that follows the legume crop because it will increase the nitrogen content in the soil and result in a better yield.

Soil conservation is an important part of sustainable agriculture. Soil conservation techniques have to be adopted to reduce soil erosion. Some of the soil conservation techniques are listed below (Adetunji et al. 2020; Koudahe et al. 2022; Crystal-Ornelas et al. 2021; Kim et al. 2020).

5.2.1 Crop Rotation

Another example from Sidéradougou, Burkina Faso, found that rotation of cowpea and groundnut with sorghum or millet resulted in an average 25% increase in cereal yield. This was attributed to the increased soil nitrogen from the legumes and the disruption of the life cycles of various parasites.

A good example of the benefits of crop rotation in Africa comes from a study in the Ivory Coast. The main staple food in this region is yams, but for four months of the year, the only cash crop available for the local farmer was soybeans. The yams are planted in ridges, and soybeans are planted between the ridges. The soybeans help prevent erosion by providing ground cover and also partially re-fertilize the ridges. This is a great example of a simple crop rotation with immediate benefits.

Crop rotation is the practice of moving different crops to different places in a systematic sequence. This has been in practice in Africa long before the chemical revolution. The primary reason for African farmers to rotate crops was to avoid pests and diseases. Crop rotation can also improve soil fertility and structure. This is very beneficial for farmers in Africa as soil fertility depletes rapidly due to the warm climate (Mpumi et al. 2020; Abbas et al. 2022; Olufolaji and Ajayi 2021).

5.2.2 Terracing

Experience from East Africa suggests that stone-built terraces have great potential and might last many years—provided they are suitably sited and constructed (Steenhuis and Gichere 1991). Although this is a step in the right direction, it is important to note that building stone terraces can actually be counterproductive for it can lead to accelerated soil erosion on hillslopes where farming is not continued, and the terrace is no longer maintained. This is where research is being conducted in Ethiopia to find an appropriate design for a given social and biophysical environment, and how communities can be involved in the selection, construction, and maintenance of various types of terraces on different land uses (Gebremedhin et al. 2006).

Terracing is the practice in which "steps" are dug into the side of a hill, which can be a very effective way of holding water in an area without it all washing away. This slow release of water can lead to plantation of different crops all year round as the moisture can be controlled depending on what the crop needs. It is also often achieved because it can convert land which is virtually useless, such as steeply sloped land, into land that can be farmed upon. This practice is commonly used in several areas around the world such as Asia and the Pacific Islands. Due to its potential and success, it is now being applied in certain areas of Africa such as Kenya and Uganda (Gashure et al. 2022; Piemontese et al. 2020; Barakagira and Ndungo 2023; Karanja et al. 2020; Kansanga et al. 2021; Kato et al. 2022).

5.2.3 Cover Cropping

In recent decades, there has been an increasing recognition that continued soil mining and degradation and declining food crop yields in Africa are not just nutrient depletion problems, but also organic matter depletion problems. Annual crop residues are an important source of organic matter in most African farming systems, but they are often insufficient to provide for both rapid mineralization and nutrient recycling and to maintain long-term improvements in soil physical and chemical properties. This is a principal cause of the low and declining labor productivity, soil moisture retention, and drought resistance of African food cropping systems, and the poor crop response to external nutrient inputs (Lal 2020; Hossain et al. 2020; Cardinael et al. 2022).

In considering how to restore soil fertility and crop productivity to nutrient-depleted soils in Africa, green-manure cover crops provide a promising and under-exploited alternative to external inputs of plant nutrients. In the first half of the twentieth century, green manures were an important feature of traditional farming systems in East and Central Africa. They were commonly used by smallholder farmers to maintain soil fertility and productivity on unfertilized and unmanured lands under subsistence food cropping. But the expansion of food production for urban markets, increased human and animal populations, and growing land scarcity led to the replacement of green manures by annual food crops and a gradual decline in their use over the past 40–50 years.

Cover cropping is another soil conservation technique which lessens the velocity of runoff, acting as a barrier. It leads to almost total soil cover, keeping it cooler in summer and warmer in winter, suppressing weeds, and increasing biological activity. Organic matter is increased, resulting in an improvement to the soil structure and an increase in water infiltration and storage. Legumes will increase soil nitrogen. A setback to cover cropping is that this system requires patience and is generally more costly in the initial years (Dossou et al. 2022; Amede et al. 2021; Bunch 2023; Yi et al. 2022; Abera and Gerkabo 2021).

5.2.4 Conservation Tillage

The amount and type of tillage permissible in conservation tillage systems vary with the specific goals and needs of producers, the crop to be grown, and the agroecological and socioeconomic conditions in which they farm. However, the primary purpose of conservation tillage in the African context is to reduce the risk of crop failure due to soil degradation from soil erosion and poor water availability. In drier regions, the primary goal is to increase the capture and use of rainfall. This is expected to be achieved in the long term by improving soil water storage potential and increasing the infiltration rate, especially on coarse-textured soils with low water-holding capacity. In higher rainfall areas, the main goal is usually to reduce soil erosion to preserve the soil resource and associated nutrient capital. This will generally enhance the sustainability of the cropping system.

Conservation tillage is the practice of preparing land for cultivation, seedbed preparation, and planting that preserves the soil and reduces erosion. Conservation tillage systems are approaches to tillage that leave at least 30% of the soil surface covered by crop residue after planting. This is different from traditional or intensive tillage, which buries the majority of the crop residue. Since it is not a specific type of tillage implement or machinery, it is a set of concepts or cropping systems that reduce soil erosion and water loss in the long term compared to the most common conventional systems. These systems have been developed in response to the high rates of soil erosion and water loss associated with conventional or intensive tillage and the need to reduce these losses for the sustainability of agriculture in Africa (Hussain et al. 2021; Diop et al. 2022; Nafi et al. 2020).

5.3 Water Management Strategies

Drip irrigation systems are state of the art methods of watering crops. They are designed to release small amounts of water directly onto the root of the plant over an extended period of time. Instead of soaking the entire field with water, the crop rows are soaked which reduces water usage and lessens the growth of weeds. This method is costly and requires a consistent water source and level of technological expertise. However, in the long term, it can be an effective way to grow high-value cash crops if the initial investment cost is feasible. An example of this can be seen in the case of smallholder chili farmers in the peri-urban area of Arusha in Tanzania. The farmers utilized a system known as the Arusha-Moshi Perimeter (AMP) to grow their crops. This was developed by the Tanga Dairy and Chillies Development Trust. The system employs gravity-fed drip irrigation, which allows minimal water usage compared to other methods and increases the output of chili from 100 kg per season to 800 kg per season (Ravisankar et al. 2021; Dickens et al. 2023).

Many of the sustainable farming practices mentioned in the previous section rely on reasonable water usage. This is what makes African farming practices different from the U.S. and Europe. Since water is a limited resource in many parts of Africa, it is essential to use what little there is in the most efficient and conservative way. While rain-fed agriculture is the primary method, the randomness and uncertainty of rainfall in semi-arid regions lead to frequent crop failures. This can be reduced by supplemental irrigation, which is becoming more available to many African farmers. There are a few very efficient and conservative methods of irrigation (Hussain et al. 2020).

5.3.1 Drip Irrigation

Drip irrigation is a kind of micro-irrigation system that has the potential to conserve water and nutrients by allowing water to drip slowly to the roots of plants, either onto the soil surface or directly onto the root zone, through a network of valves, pipes, tubing, and emitters. It is done through narrow tubes that deliver water directly to the base of the plant. It is the most efficient method of irrigation. Drip irrigation can result in healthy growth of plants, higher yield, good quality produce, and water saving. This is because it provides water to plants at a slow rate which they can absorb. The frequent application of small quantities of water is ideal for the plant where moisture stress is minimized and excess water is not left to accumulate. This is in comparison to surface irrigation where more water is applied but much is lost through runoff, evaporation, and deep drainage. It also avoids wetting the plant foliage, reducing the risk of disease. A well-designed and maintained drip system eliminates overspray, runoff, erosion, compaction, and loss of nutrients or chemicals from the crop root zone. This results in the protection of the environment.

5.3.2 Rainwater Harvesting

Smallholder farmers typically depend on rainfed agriculture, an undertaking fraught with uncertainty. Erratic rainfall is normal in many areas of Africa, where the wet season is brief and often follows a prolonged dry spell. Even in areas with adequate rainfall, frequent heavy rainfall events can result in serious soil erosion and loss of water via surface runoff, adversely affecting soil moisture retention and crop productivity (Sullivan 2003). Uncertain rainfall patterns can also lead to poor timing and over-dependence on rainfall, increasing the risk of crop failure. Rainwater harvesting is a simple, low-cost solution to alleviating these problems that is especially applicable to regions with low groundwater availability. It is a way to capture, store and use rainwater, providing a reliable source of water for crop and livestock production and alleviating reliance on erratic and insufficient rainfall. Stored rainwater can be used as a buffer, providing small-scale farmers with a means to "insure" against the risk of crop failure when rainfall is scarce.

This has been highlighted in a report by the International Water Management Institute on a study in Makanya Village, Tanzania, comparing rainwater harvesting with other water management practices for dryland maize farming. The study found that due to decreased and uncertain rainfall, traditional water management practices relying on seasonally farmed river sources were no longer a viable option for many farmers. The most successful was a farmer-managed natural valley tank with a 22 m^3 storage capacity, yielding a net annual income ranging from US$90 to US$650, claiming its convenience and reliability allowed for more intensive cultivation with improved yield, effectively lifting small-scale farmers out of subsistence agriculture (Sango 2021).

5.3.3 Water-Efficient Crop Selection

Water-efficient crop selection simply choosing crop varieties that are better suited to the water availability of the region has great potential to make farming more sustainable. Improved drought tolerance in crops and crops that require less water than current varieties offer a potential hedge against variable rainfall and can thus reduce risk in water-limited environments. A study conducted by the Consultative Group on International Agricultural Research (CGIAR) on dry land areas in Tunisia, Morocco, Algeria, and Syria demonstrated that a switch from the traditional cereal-based cropping systems to a more diverse cropping system widened the window of opportunity for planting, resulting in an increased potential crop yield. This was due to the fact that the more diverse cropping system allowed for a better use of available moisture. In addition, the study suggested a further $137 million in benefits would be incurred in the form of reduced risk due to the variance in crops worth $4 million in each country over the duration of the project. This clearly shows that although diversification of crops may not always result in higher immediate financial gains, it is a smarter economic strategy with respect to risk management.

An easy-to-understand example of an alternative crop selection would be the recent media coverage regarding the prosperity of Chinese apple farmers compared to their Australian counterparts. Due to prolonged drought in much of Australia's arid apple-growing regions, the cost of irrigation significantly reduced the competitiveness of Australian-grown apples. In addition to this, many farmers continued to grow the same apple varieties that were traditionally grown. These US/UK varieties were not suited to the climate and were thus prone to a range of problems including poor fruit color, low sweetness, and terminal diseases. In contrast, Chinese farmers observed apple growing conditions in Australia and altered their apple variety to the NZ rose series. This was a higher quality apple that required less irrigation to grow. As a result, the Chinese apple farmers captured much of the Australian apple market and substantially increased family income. At the same time, the potential of rainfall was saved and used for other uses more critical to rural Chinese communities.

5.4 Organic Pest Control Methods

Biological control is a method used to control pests by using other life forms. It can be used in two different ways: classic and augmentative. Classic biological control is used to permanently reduce the population of a pest to a low level, preventing re-infestation. This method involves the importation and establishment of natural enemies of the pest from the pest's native land. Augmentative biological control is the repeated release of natural enemies where populations are too low to maintain their effectiveness, such as in greenhouses or on some crops. Although the effectiveness of biological control remains to be proved in Africa, it is an attractive option for small subsistence farmers. It is highly specific and can be integrated into the farming system, and it also utilizes locally available natural enemies and does not require high technology. In 1980, a program was started to control the cassava mealybug in several countries in Africa, and this led to a large number of natural enemies being imported and released into Africa. This program was hailed as a major success, and it led to biological control suppressing the population of mealybugs to a low level over most of the continent.Crop rotation is a fundamental principle of organic farming, and the idea is to grow similar groups of crops in a repeating sequence and return to the original sequence after a set period. The logic is to reduce the buildup of pests and diseases that are specific to that crop. Many African smallholder farmers will be practicing a form of crop rotation but possibly with only two different crops. There have been some remarkable results in Niger where farmers have increased the number of crops grown from two to four/five and have noticed a significant decrease in the invasion of the millet crop by the bird, as the pest does not like to feed on the roots of the other crops grown. Although none of the other crops are effective trap crops for millet, this does show how the pest population can be reduced by cultivating alternative crops. The crops in Niger were rotated based on the amount of labor required and the compatibility of the different crops with traditional farming systems.

The use of beneficial insects is quite specifically a method that smallholder farmers can use. Encouraging the natural enemies of pests can be a highly effective method for pest control. Building or protecting habitat of beneficial insects and flowering plants can provide nectar and pollen for survival and longevity of insects, which can then move to a nearby crop as a food source. This strategy has been used by maize farmers in East Africa, encouraging the stem borer predator, a wasp, to control the population of stem borers. Maize interplanted with Desmodium, a leguminous plant, has shown to significantly reduce the population of the stem borer. In another project, an area-wide conservation biological control program was launched for the control of pests in the arid and semi-arid areas of East and Southern Africa. These involved farmers managing their insect pests in an area-wide context to restore and maintain the population of natural enemies, and it was designed to fit with the existing farming systems. This project, unlike many others, was farmer-led, and this increased its chances of success and sustainability over the long term (Ratto et al. 2022; Baker et al. 2020; Hill et al. 2021).

5.4.1 Biological Control

The practice of biological control involves various natural mechanisms which limit the reproduction and destroy pests by their predators. This practice is effective and is widely used by small scale farmers as well as large scale farmers. An example from Kenya concerns the farming of maize. The maize leaf-folder pest is a serious threat to the maize crop. Studies showed that the pest can be effectively controlled by a native parasitic wasp that attacks the egg stage of the leaf-folder. Farmers can therefore improve their crop yield by not using insecticidal sprays that are harmful to the environment as well as humans. These sprays kill both harmful and beneficial insects mentioned above. By not using the insecticidal sprays, farmers can save on their production costs, and protect the environment and consumer health. In Tanzania, the push-pull method has been developed through an association between farmers, and the International Centre for Tropical Agriculture. This involves an intercropping system for maize which employs companion plants that are repellent to stemborer pests. Desmodium repels the pests away from the maize and its fodder is used for livestock. The stemborer pests are attracted to napier grass which is planted around the edge of the crops. It is then heavily infested with natural pests of stemborers, the fungi Metarhizium and Beauveria. This kills the stemborer pests and protects the maize crop.

5.4.2 Crop Rotation for Pest Management

An alternate host crop such as Napier or Sudan grass may lure the borers away from maize. By chopping the grass high (leaving approximately 20 cm), the borer eggs laid on the stalks are destroyed when the stalks are cut and the larvae that hatch are unable to reach the maize. This has the added benefit of green manure and increased soil fertility from decay of the grass roots. The third maize management option is push-pull technology, which is a repellent-trap crop system created at the International Centre of Insect Physiology and Ecology (ICIPE). This involves interplanting maize with a forage legume and a series of desmodium and napier/sudan grass hedges. The semi-aquatic desmodium secretes a chemical which is harmful to the borer, inhibits its growth, and prevents it from moving to neighboring maize. The borer is then trapped in the diverse margin habitat created by the forage and grass, away from the maize where it cannot survive. This legume-grass-maize system can be directly linked to the push-pull technology development and dissemination among smallholder farmers in Sub-Saharan Africa. With another high potential option of legume or grass-based livestock feed, the desmodium and grass hedges provide an additional source of income. This project shows how complex and interlinked multiple pest management options can be brought into a crop rotation system and utilized to target one specific pest.

Some of the most sustainable and effective controls are preventative measures, such as crop rotation. Crop rotation is the practice of growing a sequence of different types of crops in the same area across a rotation of two or more years. This implements several types of controls, as pests often are specific to a type of plant and without that plant the pest cannot survive. An example of this is the maize crop in Kenya. The maize stalk borer is a major pest in maize fields and causes loss of product and increased susceptibility to other pests and diseases. The borer has now developed a resistance to pesticides, and is best treated by planting other crops.

5.4.3 Use of Beneficial Insects

The use of beneficial insects in farming is another more traditional method that can be implemented in organic farming systems. It is non-toxic and a very effective method of controlling pests. In the past, commercial farmers using chemical controls often exacerbated pest problems through eliminating natural enemies of the pests through the use of broad-spectrum pesticides. This forced the farmers to continue using chemicals to try and control the pest outbreak. Most of the chemicals killed the pests and their natural enemies at the same time. The pests, however, often have a shorter life cycle than the natural enemies and the pests reproduce at a higher rate than the natural enemies. With the natural enemies gone, the pest population will often skyrocket out of control. This creates a situation where the farmer can spend a lot of time and money trying to keep pest populations below damaging levels and the pests still end up winning in the end. The pests will cause crop damage and loss in productivity. Insects are often the most damaging pests and their natural enemies are often other insects. Encouraging the presence of beneficial insects can be a very effective tool at keeping pest populations below damaging levels at a relatively low cost. In fact, about 11% of potential crop damage is prevented by naturally occurring predators and parasites of the pests. This is equivalent to 74 million tonnes of produce with a value of 20 billion dollars. The key to success with this method is recognition and understanding of the beneficial insects and their relationship to the pests. This will enable a farmer to take action to conserve and enhance the beneficial insects and in turn, suppress the pests. This can often involve a change in the farming system from the current chemical dependent monoculture cropping to a more complex farming system. (Sawicka and Egbuna 2020).

5.5 Agroforestry and Permaculture

5.5.1 Integration of Trees in Farming Systems

Agroforestry is a sustainable farming practice which combines tree and crop management in the same area, with the trees intentionally harmonized with the crops and/or livestock. The opportunities are almost endless; a high value hardwood timber crop could be combined with a shade-tolerant medicinal plant crop, while still allowing for a mixed orchard of fruits and nuts to be grown beneath the two. The practice recognizes the complimentary effect trees and crops can have on each other. For example, trees can reduce the impact of rainfall on crops through interception, while the increased presence of vegetation and biodiversity increases the complexity of pest life-cycles, often resulting in a natural decrease in pest populations. Trees also can increase the availability of nutrients to the crops, as many trees have deep rooting systems which can bring up nutrients from the subsoil that are not accessible to the crops. Trees can enhance smallholder farming systems in Africa by providing a wide range of forest and tree products for home-consumption and sale in local markets. Farm households can meet their basic needs for food, fuel, fodder, medicine, and materials by harvesting from their trees and agroforestry plots. Income from sale of tree products can be used to meet other household needs and/or invest in farming or non-farm enterprises. As the global market for forest and tree commodities continues to grow, smallholders who are linked to these markets have an opportunity to increase their income and improve their livelihoods. By 1988, it was estimated that over 200 million Africans were involved in the production of fuelwood and charcoal, an activity that is often carried out using unsustainable methods that degrade the natural environment and reduce the availability of these resources to future generations. Encouraging smallholders to plant trees on their farms to meet the demand for fuelwood and charcoal can help to relieve the pressure on natural forests and woodlands.

5.5.1.1 Integration of Trees in Farming Systems

Smallholder tree crop plantations. This is the one sole form of tree planting where the main objective is cash income from tree crops. Smallholder tree plantations are a response to the declining access of the poor to common property resources and the increasing fragmentation of farmland.

Silvopastoral systems. This is an arrangement where trees are grown in pastures. It is of particular significance to livestock keepers, who represent 600 million of the world's poor and who are concentrated in the tropics. Like intercropping, silvopastoral systems can help meet the multiple objectives of food security, cash income, and environmental sustainability.

Agroforestry fallows: Intercropped with food crops. In this arrangement, trees are grown in the same field as food crops. The trees may be planted at the same time as the food crops or sown amongst established food crops. The main purpose of intercropping is to obtain the benefits of trees and food crops together, rather than one at the expense of the other.

Trees have been integral to farming systems in the tropics up until approximately 40 years ago. Most traditional systems allocated a large percentage of the available land to tree crops. Examples include rubber-based systems in Malaysia and shifting cultivation. Recent work by the World Agroforestry Centre has demonstrated that in many cases, tree-based systems can be more productive and/or sustainable than monocultural agriculture. There are 4 main ways in which trees can be integrated into farming systems, each with a different potential impact on food security, nutrition, cash income, and environmental sustainability.

5.5.1.2 Designing Sustainable Permaculture Gardens

Designing sustainable permaculture gardens bears little resemblance to mainstream gardening. In order to maximize the benefits, the intelligent use and conservation of energy and resources is critical. Careful and thoughtful planning is the first key to success. The efficient use of on-site resources is a fundamental principle. It entails the use of biological resources and the minimizing of all forms of waste, so that the system becomes self-sustaining. In considering a newly acquired site, or changes to an existing site, the permaculture designer will observe and interact with the land. Time spent in careful observation is time well spent, for the more information that is gathered about a site, the greater the chance for a truly sustainable design. Begin with a thoughtful and often lengthy period of producing a scaled map or plan of the site. Scaled map-making imposes a certain discipline, calling for a level of accurate observation and recording that brings important insights. Most mistakes in land management are due to a failure to observe, and the cost of ignoring or misreading the messages of the land can be both environmental and economic. On the map, all available information about the site is assembled and overlaid, from property boundaries and contours, to soil types and vegetation, to microclimate and energy flows. With the map as reference, the designer identifies the various functional elements (e.g. orchard, woodlot, vegetable garden) essential for the needs of the inhabitants and the establishment of a largely self-maintained system. The designer then makes a detailed plan for each element, progressing from generalized designs to more detailed and refined designs as each element is considered. A good design is one that is both efficient and effective. An effective design is one that achieves all of its intended purposes, whilst an efficient design is one that achieves these intended purposes with the least waste of time and energy. With every plan and element, the aim is to move towards a minimum input, maximum output system. This is done by maximizing beneficial connections between each element, so that the needs of one are filled by the products of another. High quality work is produced only with the use of high quality information.

5.5.1.3 Benefits of Agroforestry

Farmers can derive a multitude of benefits from practicing agroforestry, some of which were alluded to in the essay introduction. These include more stable and productive farming systems, increased fertility of farm land, and better protection and management of water resources. Integrated crop-livestock-tree systems have advantages over non-tree cropping systems in providing more diverse and reliable sources of food, forage, and fuel. This is mainly due to the added value of tree products and the fact that management of the total farm system is less risky when there are many sources of sustenance that can partially substitute for one another. A study by the World Agroforestry Centre in Western Kenya found that wealthier farmers planted trees to ensure greater security of future food supplies, while poorer farmers did so to meet more immediate needs. The tree to food ratios of the two groups were 28–30% and 13–20% respectively, and the most common reason for removing trees cited by both groups was to increase area under cultivation. This shows that the more economically secure farmers were willing to sacrifice immediate increases in food production for more long-term security, and that increase in farm productivity will often lead to increased planting of trees. Agroforestry is now seen as a key way to reconcile food security with environmental conservation in developing countries. By storing, recycling and better exploiting resources, agroforestry can increase total output from the land and leave the crucially important natural and built capital of farmers and their communities that underlies food security, less run down. Currently, food production systems are often short-sighted and focus on immediate returns, which can be quite destructive in the long term. The ability to use multi-purpose tree species to meet several needs at once can significantly reduce the total land area and management input required to meet those needs, thereby freeing up land and labour for other activities. This was seen in Northern Thailand where a study found that landless farmers who were given loan cows and taught improved fodder and stall feeding techniques were best able to increase their income when the program was supplemented with a tree planting initiative to provide low-cost and sustainable sources of feed.

5.6 Livestock Management Practices

Silvopasture—A sustainable agroforestry practice that involves the intentional combination of trees, forage plants, and livestock in a single management unit. The trees are managed and pruned to provide a high protein food source for the livestock while providing crucial shade and shelter. Before, judicious use of trees provides shelter, reducing the need for artificial shelters whose construction often leads to destructive land clearing and streamside erosion. Women collecting firewood. The increased mobility of livestock in silvopastoral systems often leads to less soil compaction and lower overgrazing damage. In the long term, animal health may be drastically improved due to the reduction of internal parasite loads from prolonged exposure to sunny pastures, leading to decreased reliance on chemical deworming and less impact on the environment.

Rotational grazing—The practice of regularly moving livestock from one field to another to allow the first one to rest and rejuvenate. Large ranches and small subsistence farms can use this method, which can improve plant and soil health when managed well. Benefits include more days of forage growth, increased stocking rates, and improved livestock performance because of higher quality forage. Increases in carrying capacity (the number of animals the farm can support) can eliminate the need for the clearing of new pastures. This, in turn, saves marginal lands from being exploited. The effects of grazing on streamside vegetation can be greatly reduced with rotational grazing. Wildlife habitat, especially for ground nesters, can be improved by varying the height of residual grass.

5.6.1 Rotational Grazing

When compared to set stocking, rotational grazing has been shown to have a number of positive effects. By giving forage plants a chance to rest and recover, root depths and the plant's overall vitality is increased. This has obvious benefits in plant persistence and production as well as plant palatability. The increase in root depth aids in preventing soil erosion and increases the forage plant's ability to access soil moisture and nutrients, which is particularly important in the drier seasons. With more effectively utilized forage, there is a decrease in supplementary feeding costs and/or an increase in livestock performance. With healthier plants, weed growth is better controlled and the general sward structure improves. Overall, rotational grazing can increase the carrying capacity, productivity, quality, and sustainability of a given area of land. A study from Mali comparing rotational grazing with set stocking of cattle found that "Yield of standing coarse forage was more than doubled under rotation relative to continuous grazing. High-quality forage species production was improved 5 to 10-fold under rotation."

Rotational grazing is one of the better-known land management practices. It involves moving livestock to different pastures on a regular basis. This can vary from a system of only 2 paddocks that are alternated with a rest period, to larger numbers of paddocks with more frequent moves. It is all a matter of degree. The rest period is the key to the success of the system. By resting each paddock for a long period, usually at least 6 weeks, it gives the forage plants a chance to recover and regrow to an optimal level before being revisited by grazing livestock. Rest periods can also be varied throughout the year to match forage growth patterns. Longer rests in the less productive, drier/colder season and shorter rests at the peak of the growing period. This system is easily applied in a range of scales to all livestock types and can be used in conjunction with other land management practices.

5.6.2 Silvopasture

This land use system offers the advantages of a sustainable land management strategy, an effective approach to land use, and a feasible option for land users to attain their socio-economic objectives. Due to its complexity in system layout, it may not always be the first choice for the pastoralist with land access difficulties. Yet, for those who explore the benefits silvopastoral systems have over other systems, it offers "an attractive option with potential for multifunctional agriculture and sustainable rural development."

The term "silvopasture" is a rather complex word for a simple concept and is widely regarded as the most efficient and beneficial livestock management system worldwide. Silvopasture can be defined as a form of agroforestry that involves an intentional integration of trees and grazing animals on the same land. The first form of silvopasture is practiced by allowing trees, either natural or plantation grown, to invade a current pasture land. The secondary method is the planting of forages or pasture in a wooded area and then managing the system for the two to complement each other. Yet, regardless of the method, the benefits of silvopasture are far-reaching and it is a practice that is often described as letting the land work for the people.

5.6.3 Improved Animal Housing

Improved housing can increase the health, productivity, and longevity of livestock. Many traditional livestock houses in developing countries are dark, poorly ventilated, damp, overcrowded, and insanitary. For the past 40 years, researchers have been researching the needs of different livestock species and the most appropriate ways to house them, mainly focusing on the requirements of smallholder farmers. This has resulted in a wealth of resources. Research has been substantial for poultry and pigs, with work on housing for dairy and beef cattle, sheep, and goats. From the document entitled "Design of Livestock Shelters in Sub-Saharan Africa," "A well-designed livestock shelter should provide an appropriate microenvironment for the animal. This means it should protect the animal from extremes of adverse climatic conditions, such as cold, wind, heat, and rain, but should not be so unventilated as to cause heat stress." Any improvements must be practical, achievable, and affordable. Simply tweaking the design of existing housing may be sufficient, or there may be the necessity for more radical modifications. An example of the latter is changing from a free-range scavenging system for poultry to a semi-intensive system. This will require any change to be introduced, ensuring the welfare and productivity of the birds is maintained, and that it is acceptable to the farmer, otherwise it may not last. Buildings offer the best protection, but since most small stock are kept for some time under stall feeding systems, an understanding of how to manipulate tree/shrub canopies to protect from rain or sun is very valuable. The type of flock or herd and the duration of stall feeding will affect the stocking rate of an area and the degree it can be manipulated without causing degradation. This will also determine the type of shelter to be constructed, whether temporary or more permanent. An example of this can be found in work to design simple chicken coops which have a roof of an appropriate slope for water run-off and can easily be moved, improving bird and farmer comfort compared to the traditional mud and wattle hen house. In the implementation of improvements, a considerable technical support is needed to catalyze change, both at farmer and artisan level in the case of building construction. It is often difficult to convince farmers in high-risk, high-gain situations to invest in shelter improvement as it often competes for resources with other options. This has led to experiences of housing provision as a package in relief interventions, to save animals and protect the rebuilding of livelihood within communities going through a crisis, for which field experiences are available from several farmers.

5.7 Sustainable Crop Production Techniques

Agroecology principles depend on plant biology and ecosystem ecology. An understanding of ecological and biological processes is crucial for the development of agroecological technologies. These technologies are designed to improve the sustainability of the cropping system by mimicking processes that occur in natural ecosystems. For example, diversification of the farming system seeks to develop polyculture systems that can provide the farmer with more diverse products and also serve to enhance the stability of the system. Polyculture systems can be developed through planting cover crops and intercropping cash crops; these practices can also enhance biological processes such as nitrogen fixation and the management of soil pests. Reducing the reliance on external inputs, for example the use of legume crops to provide nitrogen for subsequent crops, can mimic natural ecosystem processes and lead to more sustainable crop productivity. By understanding the complex interactions between pests and their natural enemies, farmers can design pest management strategies that do not rely on the continual use of pesticides, which often do more harm than good.

5.7.1 Agroecology Principles

Agroecology is the application of ecological science to the study, design, and management of sustainable agroecosystems. It encompasses the structure and function of plant and animal communities, and the dynamics of the environment in which they are located. It has strong connections to the sustainability of farming. Agroecology principles are used to outline the way in which farming systems need to be managed to ensure that the ecosystem related to the farm is kept in balance. This can often refer to the creation of new ecosystems or the mimicry of old native or modified ecosystems.

There are a number of principles that can be used to start turning our modern unsustainable farming systems into sustainable ones. These principles will vary depending on the type of ecosystem/farming system already in place, as well as the social and economic situation of the people that manage it. But to achieve sustainable farming systems, it is likely that all these principles will need to be used at some point.

5.7.2 Conservation Agriculture

Conservation agriculture (CA), which emerged in the mid-twentieth century, has evolved in response to the costly and irreversible failures of past soil management techniques. These had brought short-term increases in crop production at the long-term expense of environmental and resource degradation. It is based on the three linked principles of continuous minimum mechanical soil disturbance, permanent organic soil cover, and diversification of crop species grown in sequences and/or associations.

Minimum tillage systems, involving zero or no-till seeders and planters, have been found to offer benefits in terms of reducing labor and energy needs as well as by promoting soil C accumulation compared with traditional moldboard tillage.

In comparison to sole reliance upon mineral nitrogen, the leguminous under sown or cover crops used in CA systems can offer economic advantages and increased flexibility of input compared with agrochemical N sources, as well as through the additional benefits of providing organic soil cover and promoting C sequestration in the roots as well as the soil of the subsequent cash crop. The introduction of above-ground crop diversity can benefit the agroecosystem through the disruption of herbivorous pest lifecycles and decreased pest and disease incidence in the cash crop. This has led to successful development of CA systems in the humid temperate regions of North and South America involving complex rotations of several crop and cover crop species.

5.7.3 Use of Organic Fertilizers

Organic fertilizers are claimed to improve soil structure, increase soil fertility, and have a faster and longer-lasting effect than inorganic fertilizers. They also improve the water storage capacity of the soil and contribute to the control of plant diseases. There are four main types of organic fertilizer used in Africa: animal and green manure. Both involve the decomposition of organic matter and are cost-effective due to the fact that farmers can produce their own fertilizer. Farmyard manure is most beneficial to smallholder farmers because they have a ready supply of dung from their animals. However, there is a wastage of nutrients due to rain washing them out. Steps must be taken to prevent this from occurring. A study by the Kenya Soil Fertility Consortium (2006) on the use of farmyard manure in the Central and Western regions of Kenya found that it significantly increased maize yields compared to crops using inorganic fertilizer and also improved the physical and biological properties of the soil.

5.8 Community-Based Farming Initiatives

Unlike farmer field schools, cooperative farming models can involve many different types of projects. The bottom line is that they involve joint effort in one form or another. An interesting example of a small-scale cooperative project is in the Kaffrine region of Senegal where farmers came together and contributed to a community "stockage" with the aim of collectively improving food security. At the village of Keur Mame Mor Ndiaye, mixed cropping of the millet species Pennisetum glaucum (pearl millet) and Cenchrus ciliaris (a type of drought-resistant grass often grown for forage) has been set up as a rotational system to replenish soil fertility. What these examples have in common is an aim to share knowledge, and there are many platforms for knowledge sharing amongst small-scale farmers in Africa.

Farmer field schools are just one example of an initiative to make a positive impact on agriculture. The approach involves a group of farmers setting up a learning group to observe and test new methods on selected areas of their farms. The "school" mostly involves taking turns visiting the respective farms to observe and learn from the field trials, but also allocates some time for more formalized training. In Western Kenya, young farmers were taught an integrated approach to managing nematodes (a pest which attacks the roots, which is a big problem in the highly populated western districts) which involved a community effort for maximum impact. If it can be initiated successfully, it is an approach which can be self-sustaining. In addition, the fresh influx of new knowledge can revitalize a community's attitude to farming and rural development. A good variety of sustainable farming practices can be taught through FFS. In South Africa, for example, small groups of smallholder farmers in KwaZulu Natal were taught methods to improve the yield and quality of their crops, which was largely a response to widespread hunger and malnutrition in the area. This was achieved in such a way that it will benefit the environment as well as the farmers.

5.8.1 Farmer Field Schools

In response to a clear perceived need for more directed agricultural training in many rural areas, Farmer Field Schools (FFS) have emerged in recent years as a tightly structured, well-defined method for training groups of small farmers in better farm methods. The strategy was first developed and tested in the late 1980s and early 1990s by a group of Indonesian integrated pest management (IPM) specialists, in response to a perceived need for better strategies for training farmers in IPM methods. FFS has since been embraced as a model for agricultural extension—that is, training of farmers in new methods—and has been widely tested and adapted in a variety of cultural and agroecological contexts. In Sub-Saharan Africa, FFS has been widely tested in contexts, with varying success, often depending on the degree of change and intensity with which the model has been applied. The basic methodology of FFS as it has been adapted to African contexts is that farmers learn by doing—more specifically by conducting experiments in their own fields, analyses and make decisions. Its main goal is to help producers discover, through scientific methods, the best ways to improve their yields and the quality of their products. This is to be achieved by collective action of the group. Its underlying premise is that sustainable and cost-effective gains in agricultural production and income can be best achieved by taking into account the complex and dynamic agro-ecological conditions, and thus investing in local human resources. Farmers Field Schools can last from one cropping season to as long as years depending on the lesson being taught. The lessons also vary from agriculture, pest control, and marketing. The skill is not intended to be transferred to the farmers, but learned within the context of farmers discovering and creating their own solutions to their problems. The assumption is that learning how to learn is more important than what to learn.

5.8.2 Cooperative Farming Models

Membership in a cooperative has enabled women farmers to pool resources (such as land or labor) to increase their agricultural productivity, to access credit or agricultural inputs which they may not be able to obtain on their own, and to market their products more effectively for better returns. In Cameroon, women vegetable producers who joined vegetable marketing cooperatives had a 22% increase in annual income compared to non-members. Cooperatives frequently engage in training activities to build member skills and knowledge and have gradually begun to incorporate a stronger gender focus into their programming. By joining a cooperative, women have greater access to decision-making forums where their voices can be heard and where they can advocate for changes that promote gender equality within the cooperative and in the community in general. In Uganda, women who were involved in multipurpose cooperatives were found to have increased their participation in community development activities five-fold.

In severe rural poverty-stricken areas of Africa, smallholder farmers can barely produce enough food to feed their families, let alone reap a profit that can improve their standard of living. A significant number of these farmers are women, who are the primary tillers of the soil in sub-Saharan Africa. Farming as a woman often means dual responsibilities of working in the fields and caring for the family, with little control over resources. Women are disproportionately in the bottom ranks of agricultural wage employment, and their issues often go unaddressed. The cooperative farming model has had a profound effect on changing this dynamic for women farmers. Cooperatives are collective organizations owned and operated by the members they serve. They range in size from small farmer groups to multi-million-dollar businesses.

5.8.3 Knowledge Sharing Platforms

An example of a successful knowledge-sharing platform in the context of sustainable agriculture in southern Africa is the Farmer Research Committee (FRC) approach. This is an initiative led by national agricultural research and extension systems (NARES) and NGOs whereby smallholder farmers are provided with a forum to identify their own research priorities and develop solutions in partnership with researchers. By engaging in participatory research initiatives, farmers gain access to new information and technologies, which can improve their production, whereas researchers gain understanding of the localized problems experienced by farmers and new innovations can be developed. This platform provides a two-way learning process for both farmers and researchers. An evaluation of FRC in Zimbabwe has shown that the platform has facilitated enhanced institutional linkages between farmers and researchers, novel technology development, and increased productivity on the farm.

Knowledge sharing platforms are essentially designed to enable interaction and networking among people to promote sharing, co-creation, and dissemination of knowledge. Such platforms can assume various forms, including conferences, meetings, workshops, forums, etc., which can be either virtual or physical. Ideally, knowledge sharing platforms can be an effective means of enabling farmers to identify and seek solutions to their everyday production constraints through interaction with their peers and other stakeholders in the research and development domain.

5.9 Climate Change Adaptation in Agriculture

Climate change will greatly impact African agriculture. Changes in temperatures and rainfall will result in possible yield reductions and modification of suitable land for specific crops. Agriculture, particularly in semi-arid and arid areas, faces an enormous task in adapting to climate change. However, the projected impacts of climate change on Africa and the fact that the continent is so poorly resourced in terms of being able to respond heightens the importance of African agriculture adapting to climate change. In a report by the Consultative Group on International Agricultural Research (CGIAR) climate change research program, the severity of the impacts of climate change on agriculture in Africa was stressed, despite the continent only contributing 3.6% of global greenhouse gas emissions. The CGIAR believes that the key for success in African agriculture to be able to adapt is in developing three traits: more resilient crop varieties, diversification of farming systems with self-sustaining practices, and policies that are supported by research and development. This is reflected in realizing the need to bring about a green revolution in Africa that must be tailored to the continent's special needs. This section will briefly explore some of the best ways African agriculture can adapt to climate change.

5.9.1 Resilient Crop Varieties

Resilient crop varieties can be identified, propagated, and used to great benefit for coping with crises such as the current COVID-19 pandemic. A resilient crop variety is one that continues to meet the requirements of producers, consumers, and the wider society, despite a rapidly changing and challenging environment. Under the pressures of a rapidly changing climate and a more variable climate, the abiotic and biotic stresses affecting crop production are bound to alter in nature, posing new challenges to food security. For example, the increased incidence of pests and diseases due to climate change will mean that crop varieties which were once resistant may no longer be so. In some instances, varieties of the same crop that were once grown successfully will no longer be a viable option. Therefore, pre-empting the challenges and identifying the traits needed in future crop varieties will be a key part of climate change adaptation in agriculture. An illustration of this would be the anticipated increased UV radiation at higher latitudes. This is likely to increase the incidence of skin cancers in fair-skinned populations and will necessitate behavior change and also medical and biological adaptations. Similarly, it will also cause an increase in skin cancers of livestock with pale skin and non-pigmented areas. This, in turn, will necessitate new methods of prevention and treatment for animal cancers and also breeding strategies aimed at increasing the pigmentation of animal skin in susceptible areas. For these anticipated future changes, it will be necessary to begin now to identify the trait changes and new varieties required, and where possible make a start on breeding and testing. This will ensure a sustainable food supply, but often the desired traits will take many years to breed into crop varieties. In the short term, it may be possible to select or induce the required traits in existing varieties. An example of pre-emptive breeding and testing of new varieties is the work carried out to identify varieties of rice and wheat capable of providing stable yields in environments of increased CO_2, which inhibit the growth of plants utilizing the C3 pathway (Zsögön et al. 2022; Shahzad et al. 2021).

5.9.2 Agroforestry for Climate Resilience

Agroforestry is a land use management system in which trees and shrubs are grown around or among crops or pastureland. It combines agricultural and forestry technologies to create more diverse, productive, profitable, healthy, and sustainable land-use systems. Trees are often absent from or only an afterthought in agricultural systems, yet they can provide multiple benefits over several decades. Agroforestry increases above- and below-ground biodiversity as well as soil biological activity. These, in turn, lead to increased productivity and sustainability of the total system. Shelterbelts, living fences, and other agroforestry practices can protect crops and livestock from damaging winds. Integrated crop/livestock/tree systems can provide a hedge against the risk of complete crop failure in that some sustenance will be provided by the cattle and/or trees, thereby spreading risk over a number of enterprises (diversification). This is an important component in coping with unpredictable and often harmful variation in climatic conditions.

Agroforestry systems also provide direct protection for trees and provide diverse and frequent choices for farmers. Mixed systems can provide yields on a very sustainable basis, thereby reducing the need for often environmentally costly land expansion (this is the aim of the Plant a Billion Trees campaign). These trees are often planted on degraded lands, and research on revegetation and regeneration across the tropics has global relevance for applying agroforestry to cope with climate change.

5.9.3 Climate-Smart Farming Practices

A key step towards climate-smart farming is sustainable land management. Most food production in the developing world is rainfed agriculture, often on marginal lands. The intensification of rainfall leads to soil erosion, lower soil fertility, and nutrient leaching. Unsustainable land management practices are, in turn, a driver of deforestation and wetland loss. In an effort to maintain a balance between food security and environmental conservation, it is pivotal for agriculture to shift to more sustainable land management. This would involve the improvement of soil quality, terracing on sloping lands, and reducing the pressure on highland and forest ecosystems. However, this change will involve transition costs and the need for a policy-level enabling environment. In Nigeria, for example, a move from shifting cultivation to more intensive cropping systems has been proposed, but a lack of infrastructure and accessibility to suitable land may hinder this move.

Climate-smart farming integrates adaptation and mitigation in agricultural development. The productivity of climate-smart farming can be assessed through three main parameters, which are: (i) increased productivity—improving food security and farmers' income, (ii) enhanced resilience—reducing vulnerability to climate change, and (iii) reduced greenhouse gas emissions—which contribute to mitigation. Although the potential benefits are high, adoption by farmers is still low due to a wide array of constraints.

5.10 Case Studies of Successful Sustainable Farming Projects

Project B is based in Bokoji, a village in southern Ethiopia. The people in the area are marginalized subsistence farmers that have insufficient land to feed their families, and deforestation and overgrazing has left the environment degraded. A non-government organization named SustainE founded a project in the area that worked with the farmers towards sustainable agriculture, improving the livelihood of the people and rehabilitating the environment. The project encompassed a number of aims but centered around agroecology. The project consisted of two phases, the first was participatory rural appraisal (PRA). This involved group discussions with the farmers to identify the problems, causes and possible solutions in the community. This gave the farmers a sense of ownership of the project and has instilled continued enthusiasm and participation of the farmers. The second phase was capacity building, or the training of the farmers. The main activities were training on soil and water conservation and agroecology. The training focused on conserving the environment and rehabilitating the degraded land. The farmers were trained that with each activity they do they should consider the effects on the environment, and that a healthy environment would bring long term prosperity. This was a huge success and to this day the farmer's still subscribe to the project aims.

Project A took place in Tharaka, located in Kenya, and is a successful example of sustainable farming. This dry region is susceptible to drought and famines are common. The Meru Central Dairy Union began work in the area in 1999 with the intention of improving milk production and quality, that would in turn raise the income of the people in the area. The Union brought in dairy cows from other areas of Kenya that were already accustomed to arid conditions. Then an exchange program was set up, between the people, farmers were provided with cows, and in exchange a few months later would provide a calf back to the program. The cows have been very productive in the area and the subsequent cross breeding with local breeds has led to improved milk production of all cows in the area. Dairy farmers in the region have also been provided with training towards organic farming practice. Work has shown that in the long run this is best way for small scale farmers to improve their income. An Organic farming project was initiated to produce Napier grass with legume forage that had higher nutrient levels and improved soil retention, as well as crop farmer training on production of feed. This meant that milk production had quality and quantity increases and it raised the income of the farmers, an average of 30% increase. This has led to increased school attendance of the children, due to increased ability of the parents to pay fees, and better living conditions. Overall the quality of life in the area has improved.

5.10.1 Project A: Sustainable Farming in Kenya

The project was conducted under varying socio-political and agroecological conditions and involved a sequence of rapid appraisals and long-term studies in clusters of farm households. The first phase in 1996 was in Tea Zones where farming and environment are in a state of rapid change, the second phase was in Sirwo Mixed Farm System which was a study of soil and vegetation dynamics and farmer responses, and the third phase was in Mauche Agroforestry and was predominantly a social science study. During the project, researcher and farmer interactions and feedback of results were promoted through farmer workshop activities. Farmer resource networks were also established in collaboration with KARI to utilize present and historical farming systems and farmer responses to treatment effects for the benefit of farmers and future research.

Collaborative research between a Danish and Kenyan institution has been conducted on sustainable farming practices in Western Kenya from 1996 to 2003. The project, integrating natural and social science, is known as "Drawing Lessons for Integrated Assessment and Sustainable Development Initiatives (DELSA)". This research aimed to provide insights into the selective pressures that shape the natural resource base of farming systems, and indicate options for sustainable development. The whole project was based on the premise that interdisciplinary analysis is a necessary component of problem solving and participatory research is vital for the validation of research results as well as to empower resource users to manage their environment.

5.10.2 Project B: Organic Farming in Ethiopia

I. R8761: Human Resource Development and Curricula Review

This operated from 1998 to 2003 and had three components, working with 3 universities:University staff were sent to the UK for MSc and PhD degrees, particularly (but not exclusively) in soil science, plant breeding, agricultural economics and agricultural education. This was beneficial to UK institutions which found Ethiopian students among the best in Africa.

Short course provision for both undergraduates and later appointees.

Review of undergraduate curricula.

Components I la and I b were particularly well received. Over the years about 50 people were sent to the UK and about 3000 took short courses. In component II we received high quality feedback and the training itself led to job creation. Cost effectiveness was high and activities were well integrated with UK partners.

This project began organically in 1992, with the first training workshops at the then Alemaya College of Agriculture. Prior to ESSP, international inputs into support for Ethiopian agricultural education in the previous 15 years had been inert. Competitive scholarships for higher degrees to a few individuals, purchase of equipment and provision of block funding for departmental improvement, may have improved the performance of individual faculty but had little impact on the institution as a whole. This situation changed in 1998 with three projects, strengthening faculties in three universities (Alemaya, Hawassa and Haramaya).

5.10.3 Project C: Agroecology in South Africa

Agroecology is a holistic science based on the application of ecological principles to the design and management of sustainable agroecosystems. During the early 1990s, the Ford Foundation funded an effort to research the possibilities for sustainable agriculture in South Africa. The initiative aimed at increasing food security for small scale and resource poor farmers in ways that would uphold land conservation and economic equity. The program involved training field workers and agricultural extension officers in agroecology, and forming partnerships between NGOs and farmers to establish learning sites and agricultural experiments. This was called the Farming Systems Research Unit (FSRU). One of the partnerships that evolved from the FSRU was with community members in the Kransbos area of the Eastern Cape. This was a participatory research project where the farmers and the researchers learned and designed experiments together. The focus of this initiative was the identification of sustainable land management practices that would result in improved carrying capacity of communal lands. A secondary goal was the identification of uses for indigenous plants that could generate income, without compromising land conservation. During this period, the civil society organizations assisted in the advocacy against the GEAR (growth employment and redistribution) economic policy which undermined the transformation to a more just and equitable society. A recent collaboration between the University of Cape Town and the NGO Resource Africa has sought to use farmer-to-farmer agroecology knowledge exchanges as a way of enhancing natural resource management in the Pondoland area. This project recognizes that knowledge systems determine practices and it is necessary for farmers to lead their own initiatives of natural resource management (Thornton et al. 2022; Ulian et al. 2020).

5.11 Challenges and Barriers to Sustainable Farming in Africa

The majority of the poor lack access to land and other resources, and this problem is likely to worsen as growing populations seek to make a living from farming. Many have no option but to farm marginal lands, posing serious risks of resource degradation. It is essential to increase investment in agriculture to provide the necessary resources and to promote the development and adoption of technologies that will help farmers to use these resources more efficiently. Investment is needed not only to raise productivity on farms, but also to better integrate small-scale farmers into the market economy.

This will involve improving the linkages between farmers and input suppliers, output markets, and credit and savings facilities. But changing the face of African agriculture will not happen without a significant change in the way that policy makers and planners view the small-scale farming sector.

5.11.1 Limited Access to Resources

Farmers in dryland areas are at a distinct disadvantage in their quest for sustainable agriculture. A shortage of water inhibits the production of biomass and crop residues that are essential to soil fertility improvements. This in turn affects the ability of farmers to increase food production, leading to a downward spiral of declining sustainability. Many are forced to over cultivate their land further degrading its quality. A lack of resources also affects the ability of farmers to adopt new technologies perceived as being risky. If a farmer can only just grow enough food to survive, investing in a new technology that only increases long term production doesn't make sense if it reduces their ability to eat in the short term. Although access to relatively cheap migrant labour can reduce some of these constraints, the artisan knowledge required to implement SLM practices may be lost. This is because many traditional water harvesting and soil conservation practices that were sustainable have been replaced by modern methods that are unsustainable with the limited resources available. This effectively locks the farmer into a cycle of declining sustainability. High upfront costs, for example land levelling required for the construction of water retention ditches or conservation furrows, can make the implementation of these practices prohibitive. This is especially true if the land is only being leased as is often the case with women farmers. Usually, women have less access to resources and may also have less land tenure security.

5.11.2 Lack of Awareness and Education

The lack of information or education on sustainable farming is a critical difficulty for farmers in Africa. This calls for discouraging techniques since farmers are not aware of the effects of their farming practices. Comprehensive research on sustainable agriculture information and learning programs, both within and outside Africa, is limited. This situation is further compounded by the generally low levels of literacy among the older rural farmers who often have the greatest stake in sustainable agriculture. The result is that knowledge about sustainable agriculture by farmers is often sketchy and fragmented. Unfortunately, where the hazards involved are not clearly understood, the advantages of new practices or technologies are not seen as outweighing the costs. A study of Malawian smallholders identified that the concept of conservation farming, which in the Malawian context involves minimal soil disturbance, permanent organic soil cover, and crop rotation, was interpreted narrowly in terms of the permanent organic soil cover, with little changes to the traditional practice of hand hoe tillage and without the crop rotation.

5.11.3 Policy and Institutional Constraints

Like other parts of the world, African governments have signed and ratified international agreements and treaties, which include provisions to increase sustainable development and natural resource management. The failure to translate these formal commitments into tangible action and results can be attributed to a combination of political, economic, social, and institutional factors. This is despite the fact that African countries have a mix of traditional and modern institutions, some of which offer valuable lessons regarding sustainable development. Unfortunately, most African countries have colonial heritage, where policies were skewed to serve the colonial administration and generally led to inequitable access and benefits to resources, particularly in rural areas. The legacy of colonialism often weighs heavily upon their ability to develop effective policies for sustainable natural resource management that will redress the imbalances of the past, primarily because the entrenched patronage systems and structures of control and exploitation by a small elite have suited post-colonial African ruling classes during the last four decades.

In today's rapidly globalising world, African countries are also often on the periphery as they are unable to actively participate in international negotiations on various global agreements, including vital issues such as climate change. This means that whilst many international agreements are designed to be universal, they may not necessarily take into account the special circumstances in Africa or have sufficient provisions for technical or financial assistance for participation and implementation. This can often contradict with the sovereign power of a nation and lead to policy-making conditions that are more likely to further marginalise certain sectors of society and constrain sustainable management of resources. In the global institutions themselves, there are often conflicting agendas and conditionalities of aid and loans, which can further pressure African countries to exploit resources unsustainably in order to fulfil short-term development objectives.

5.12 Future Prospects and Recommendations

The ability to advocate and negotiate policy change will depend on the capacity of different stakeholder groups, and there is solid evidence that investing in the empowerment, knowledge, and skills of resource-poor farmers can lead to improvements in their livelihoods and food security while preserving the natural resource base. Given this, investment in human and social capital needs to be a central component of efforts to develop sustainable farming practices. This can involve new forms of education and extension provision, farmer-led research and innovation, greater access to information and rural advisory services, and capacity building for various forms of collective action among farmers. An increasingly important issue is the empowerment of women who are a major but often neglected stakeholder in agricultural and rural development. Women tend to be more resource management-oriented than men, and specific efforts to promote their access to resources and participation in decision-making will likely have multiple benefits for household food security and the environment.

Successful scaling up sustainable farming practices will depend on the type of policy environment that creates incentives for change, provides public goods such as rural extension and research services, and also levels the playing field between sustainable and conventional alternatives. Drawing upon global and local experience, there is a need to move beyond isolated sustainable farming projects towards the integration of principles and practices of sustainable farming into mainstream agricultural policy and program development. This will involve the creation of partnerships with a wider set of stakeholders, advocacy for policy change at various levels, and the involvement of farmers themselves in policy and program formulation. Although the political contexts and processes of policy change will vary from country to country, there are instances where successful sustainable farming projects have seen national policy impacts.

Adopting sustainable farming practices will require a well-coordinated strategy across several sectors. Hence, there is a need to focus on the broader development context within which agricultural development occurs and a need to develop a new generation of sustainable farming practices in the context of global change. First and foremost, there is a need to scale up successful sustainable farming practices from local or pilot level to more widespread adoption. One example is the system of rice intensification (SRI), which began in Madagascar and has now spread to over 15 other countries in Asia and Africa. SRI is a good example of a new generation sustainable farming practice that has been developed via farmer-led innovation. Another example is conservation agriculture (CA), which is gaining growing acceptance in the developing world and has the potential to break the cycle of resource run down in fragile and marginal environments. For such practices to reach their full potential, there will need to be increased investment in basic and adaptive research, stronger linkages between public and private sector partners, and the development of inclusive value chains for smallholder producers to access essential inputs and adopt new innovations.

5.12.1 Scaling Up Sustainable Farming Practices

Scaling up successful practices to the regional or national level is a complex and long-term process, faced with numerous obstacles including resource scarcity, risk and limited access to support networks. The complexity of the process is being recognized by a shift in development assistance towards "participatory", "learning by doing" approaches, as opposed to the identification of best practice and its simple transfer to farmers. For example, in a study of soil fertility management in nine countries across east and southern Africa, it was found that best results were achieved through repetitive community meetings and field trials, and the worst from the dissemination of pre-packaged technologies. The time and resources required to facilitate such processes at a scale relevant to millions of smallholders are however rarely committed. Efforts to facilitate such processes at a scale relevant to millions of smallholders are however rarely committed. Commercial and government-run extension services are in many countries in decline, yet support networks of rural NGOs and CBOs are still weak and reliant on external funding. Many farmers and farming households are themselves risk averse, and given the increased risk and resource demands of new practices, are unlikely to adopt them without the assurance of lasting benefits. Although many sustainable practices are cost-effective in the long-term, establishing their profitability relative to conventional practices on degraded soils with low rainfall can be difficult. An example from Zimbabwe illustrates the constraints: despite success at the individual level and high social demand for agroforestry, formal woodlot establishment has required a high level of external subsidy.

Sustainable farming practices are being adopted at the local level across the continent, yet national and international support for them has been limited. Evidence from case studies suggests that sustainable agricultural practices can be highly productive and can result in reduced pressure on fragile environments. For example, in Machakos district of Kenya, participatory research in soil and water conservation has enabled farmers to generate enough food from their unimproved farms to meet their needs, thus avoiding further encroachment into marginal lands. However, in areas of high population density, land scarcity and declining soil productivity, such local level successes are difficult to maintain.

5.12.2 Strengthening Policy Support

The situation in Africa is varied: there are some countries emerging from conflict with very poor governance and state capacity, and here the priority is to re-establish functioning states that can ensure security and deliver public goods. In other countries, the concern is that policies in various sectors are often inimical to the interests of smallholder farmers, even though official statements may assert the contrary. This may require changes in specific sectoral policies as well as more general changes in economic and institutional regimes.

There are a number of past examples in developing and industrialized countries of the ways in which policy environments can be modified to encourage changes in practice, with mixed results. Often the more prescriptive and enforcement-oriented approaches (such as environmental regulations or directives) are resisted or side-stepped and may result in perverse outcomes. In some cases, the most effective strategies have been to create the right enabling.

It is essential for activities in the first two areas that are aimed at scaling up sustainable agriculture to be accompanied by a supportive policy environment. This is because many of the sustainable practices run counter to the current economic and institutional incentives for farmers and consumers, and to be effective, alternative practices will often need public sector involvement to create supportive conditions.

5.12.3 Investing in Research and Development

In general, the productivity of African agriculture is low by global standards. However, there are significant differences in the levels of productivity between the various countries and regions of Africa. There are some pockets of African agriculture that are highly productive and are able to compete effectively on both the internal and external markets. There is often little understanding of why these islands of productivity exist and whether there are any learnings that can be transferred to other agricultural systems. There is evidence that success in these cases has been associated with higher levels of investment in R&D. However, the level of investment in R&D in African agriculture has been low and declining, both in absolute terms and relative to other regions of the world. The key challenges in investing in R&D are to convince policymakers of the importance of the issue, and to invest the available resources in ways that are relevant, effective and efficient.

This is a major challenge given the complex and diverse nature of African agriculture.In the long-term R&D is also likely to be the most cost-effective way of improving the productivity and sustainability of African agriculture. A better understanding of the biophysical and socioeconomic constraints to agriculture will help to target investments in agriculture to areas where they will have greatest impact. This will require multi-disciplinary programs that are well linked to policy and development community. R&D will also be critical in developing the technologies and management practices that can help smallholders to intensify their production in an environmentally sustainable manner. Finally, the dynamic and complex nature of African agriculture calls for ongoing monitoring and foresight so that policy can be continually adapted to the evolving circumstances (Nin-Pratt 2021; Adetutu and Ajayi 2020)

5.13 The Role of Traditional Knowledge in Agriculture and Food Security in Africa

The study is structured as follows. The aim of this issue is to recognize that traditional knowledge has played a key role in maintaining food security and how this can be supported in the future. Traditional knowledge systems in agriculture and food have an important role in maintaining human well-being by providing culturally appropriate and locally accessible food, preventing diseases, and also enhancing the quality of the environment. To introduce the importance of traditional knowledge in agriculture, the first theoretical section of the paper provides a brief overview of traditional knowledge and key work that has been done in the field. Moving to look at regional food security situation, it is clear that Africa has the highest incidence of food insecurity in the world. Food security exists when all people, at all times, have physical and economic access to sufficient, safe and nutritious food to meet their dietary needs and food preferences for an active and healthy life. This is another area where traditional knowledge has a role. The two papers focusing on case studies have shown that traditional knowledge can often be the difference between a secure and insecure food supply. This issue is very important for policy makers and the public to understand as it debunks the common myth that traditional knowledge in agriculture is backwards and unprogressive. By showing the importance of traditional knowledge in maintaining food security, it is easier to make a case for the support and continuation of traditional knowledge systems.

5.13.1 Importance of Traditional Knowledge in Agriculture

Traditional agriculture's main aim is subsistence for survival and maintenance. Regular cash cropping only takes place on the sale of surplus goods, and as a result, production is for consumption and is more often than not dependent on satisfactory rainfall. Uses of land are not to maximize profits, but to satisfy household food requirements and to ensure future production from the land. Due to this, traditional agriculture is sustainable and has a minimum negative impact on the environment. This is contrasting to modern agriculture, which is input intensive and focused on high outputs for economic gain. The results are often land degradation, and in the African context, it is often a fall into the poverty cycle for small-scale farmers.

Traditional knowledge forms the basis of indigenous communities' identities, cultures, and their survival and well-being. It is validated and evolved through generations of trial and error experimentation, and forms the foundation of the community's agricultural systems. A study carried out by the World Bank and the Food and Agriculture Organization reviewed the state of agricultural and natural resources knowledge in Africa. Its results concluded that indigenous knowledge is critical to the future well-being of any rural communities in Africa, that there is a wealth of generational knowledge that aids farmers in food production and natural resource management, and that increased food insecurity and land degradation experienced on the continent by African farmers has caused a decline in the valuable traditional knowledge because the elders are dying and the youth seek more modern lifestyles, as well as better education to overcome their situation.

5.13.2 Significance of Food Security in Africa

In Africa today, the import of guaranteeing food security to each household is up against an array of challenges, the foremost being the impact of HIV/AIDS and the stigma associated with it. This pandemic is reducing the human resources capacity for food production and the ability of households to carry out agricultural activities and provide their own food, especially in severely affected countries and areas. This, in turn, reduces the availability of food on the market and can thus disrupt the entitlements of less severe cases and those vulnerable groups not yet affected to meet their food requirements. Many will move to try and seek labor work in urban areas or other rural areas to supplement the loss of income, further depleting the human resources capacity in their own households and/or on their land to produce food. Other coping strategies employed by these affected households include, but are not limited to, exploitation of natural resources for immediate food and income which, if unsustainable, can further jeopardize future ability to produce food. A decrease in purchasing power will limit the types and amount of food which can be bought, further undermining food access and utilization. Overall household food security status will be greatly reduced, and more households will fall into the state of food insecurity. Another major barrier to food security in Africa is armed conflict, which is often linked to other socio-political problems. This can lead to the destruction of infrastructure in many areas, limiting the availability of food and access to it and seriously disrupting entitlements. People affected by conflict are generally forced to move away from their homes and agricultural lands and may become refugees or internally displaced persons. These people are entirely dependent upon uncertain food aid and are the most vulnerable group to food insecurity in host countries. Globalization or, more specifically, globalized markets have increased the availability of different types of food in urban areas and among the rich in many African countries. While this is positive for those with increased purchasing power, it has negative effects for small-scale farmers who cannot compete and may switch from production for own consumption to cash cropping and buying food, further reducing food access for themselves. Other farmers may switch to cultivation of non-traditional foods and import foods which are unpopular among the majority but more profitable. This decreases community food availability and increases consumption of non-nutritious foods (Lieber et al. 2021; Oluma et al. 2020; De Waal 2020).

5.13.3 Traditional Knowledge in Agriculture

As people are still engaged in agriculture in North America, traditional knowledge is often not utilized or is taken and altered to better serve the needs of the farmer. This trend is also beginning to take place in Africa. There are multitudes of traditional farming techniques that are being lost as African peoples begin to use modern techniques. One of the reasons modern techniques are being used is that they are often seen as being more efficient in producing food for the time and labor that is required. The quality of the food and the long-term sustainability are not considered.

Those who still utilize traditional techniques do so because they believe the quality of the food is better, there is less waste in terms of usable crop in comparison to the labor put in, and the land can be used sustainably for a long period of time. The important thing to note is that if the quality and quantity of food between the two techniques was equal, no farmer would want to spend longer hours in the field for the same amount of product. The migration to modern techniques threatens the food security in Africa by eroding away traditional techniques and often not providing an adequate replacement. An example of this is the ox-plough cultivation by the Baganda in Uganda. A study found that despite many people favoring ox-plough cultivation and it being considered the best technique for cultivating certain crops, it is being replaced by hand hoe cultivation. The reason for this shift was that hand hoe cultivation was often considered a necessary first step to adopting plough cultivation. Furthermore, hand hoe cultivation was also viewed as a symbol of a poor man's inability to own oxen. The recent wars in Uganda had a large impact on this shift. Poor men often lost their cattle to soldiers looking for food and plunder and were therefore forced to rent oxen to cultivate or cultivate by hand. The displacement of this technique was not due to the inefficiency of ox-plough cultivation, but rather outside influences and the current impact of the situation in Uganda. An inadequate replacement of technique to plough certain crops can result in lower yields, and little progress has been made in terms of plough cultivation since the displacement. This is a technique which if lost, would hinder an already poor society from increasing its food security. The Baganda must find an effective way in which to pass on plough cultivation to the next generation and try to revive its use. The study concluded that to prevent further loss of technique, promotion of plough cultivation and assurance that there are no threats to its practice are necessary. This may need to be supplemented by a system in which certain traditional techniques are protected from modernization until an adequate modern technique can be found. This will be a difficult task since it pulls the direction of farming in Africa in the opposite direction of current global trends (Britwum and Demont 2022; Palanivel and Shah 2021; Ratemo et al. 2020).

5.13.4 Traditional Farming Techniques

Traditional farming techniques in Africa are extremely diverse. Different cultures have developed an array of innovative farming systems and practices based on site-specific ecological, social, and economic factors. These systems have sustained millions of resource-poor smallholder farmers for hundreds of years. Central to these systems is the idea of "careful observation". Farmers have observed the growth patterns of different crops, they have noticed the interactions between different crops and pests, and they have learned from the successes and failures of others. This careful observation has led to the development of a set of proven farming techniques. Many of these techniques are low-input/low-risk options that are easily accessible to resource-poor farmers. These include the development of an array of planting basins, ridges, and mounds to capture and utilize rainwater, increasing infiltration and reducing soil erosion, and the use of natural vegetative and soil amendments to increase soil fertility. Animal traction is still the most sustainable form of farm power for many smallholder farmers, and a range of farming methods have been developed and diversified to suit the different agro-ecological areas. This includes mixed farming, relay cropping, and rotational grazing systems. (Giller et al. 2021)

5.13.5 Crop Diversity and Traditional Seed Saving

Crop diversity is one of the most vital aspects of traditional knowledge in agriculture (TKA). Traditional farmers of all societies have long maintained a wide diversity of crop varieties. This is not only the most obvious manifestation of genetic resources, but the most critical as well. It reflects a deep understanding of the environment, a highly developed system of natural resource management, and an interdependent relationship between humans and their environment. Where such diversity exists, it is a clear indication of continued selection and improvement of locally adapted varieties. Traditional farmers grow many different crops in the same general area. Usually, the topography of the region would be divided into a complex of microenvironments, each of which may be suitable for a specific crop. In the tropical forests and highlands, the diversity is even more pronounced. Different plant species will often be found living together in the same environment. This has resulted in an enormous variety of crop types and species. The great diversity of crops found in such areas often results in a complex of wild and domesticated species, varieties and landraces. This has occurred where farmers have continued to evolve new crop types from wild progenitors, and have maintained these with a high level of genetic integrity. Unfortunately, much of this crop diversity is now being eroded. One of the reasons that traditional farmers have been able to maintain such a high level of crop diversity is that they have developed their own seed selection and saving practices over many generations. These have often been tailored to specific varieties and environmental conditions. In many societies, crop seed and propagation material has a specific cultural and spiritual significance, and is often subject to complex taboos regarding its handling and exchange (Khoury et al. 2022; Ogunyiola et al. 2022; Labeyrie et al. 2021).

5.13.6 Natural Pest Control Methods

Natural pest control is a cornerstone of traditional agro-ecological knowledge, which seeks to work with natural processes and cycles rather than bypassing them with chemical inputs. It is intricately linked with the growth of certain crops in space and time, especially polyculture in which several crops coexist throughout the year. Mixtures of crops can confuse pests and slow the spread of pests and diseases compared to the spread of a single crop. Pests are less likely to locally extinct in a diverse ecosystem where there are alternate hosts and predators. Traditional shifting cultivation involving a mosaic of fields in different successional stages can also reduce pest problems by decreasing the amount of time a field is under the specific pest's host crop. Weeding is often an important part of pest control and can be highly specific, for example pushing weeds between the rows of maize so that the weed-feeding Striga stem-borer will be enticed to lay eggs on the Striga rather than the maize. Many societies use their detailed ecological knowledge to place certain crops in micro-ecosystems where they are less likely to be attacked by certain pests or diseases. Knowledge of an pests habits and life cycles is used to time planting so that the crop will emerge at a time when the pest is not active. Tagbanwa upland rice farmers in the Philippines plant their rice just before a known migration of rice black bug to higher elevations, exposing the rice to a shorter period of infestation. Elsewhere farmers have used knowledge of rainfall patterns to plant dryland crops in a year with lower rainfall, minimizing the risk of disease spread from a healthy crop to a neighboring wetland crop.

5.14 Traditional Knowledge in Food Security

Indigenous food systems are an integral part of a community's identity. They give people a sense of who they are, where they come from, and where they are going. It is rooted in culture and place and linked to issues of empowerment and control, as well as to health and well-being. One of the more utilitarian aspects of indigenous food systems is their role in food security. Food security can be described as access at all times to sufficient, safe, nutritious food to maintain a healthy and active life. Indigenous food systems make an important contribution to food security for many communities, even though they are rarely reflected in the mainstream discourse about sustainable food systems and development. They are often perceived as backward and underdeveloped forms of agriculture and a barrier to food security. This is an erroneous perception which is based on a lack of understanding or appreciation of what indigenous food systems entail and the role they play for many communities, and a bias towards industrial agriculture and food systems as the only way forward for "developing". Although food security is changing for many communities and issues of access to land and resources, autonomy and support for young farmers are also important, for the purpose of this document we will be examining food security as it applies to subsistence agriculture and consumption, particularly for rural communities. Traditional food systems are place-based and socially embedded with a strong knowledge component which encompasses the growing, gathering, raising, hunting, and fishing of food as well as its preparation and consumption. This knowledge is often unique to the ecosystem and environment of its origin and is transmitted through generations of trial and observation, error and success. In the context of indigenous peoples, food and cultural identity are intertwined and central to their survival, particularly for those who have been displaced from their homelands and traditions. For such communities, maintaining access to traditional foods and the knowledge of how to procure and prepare them can be a means of resistance and resilience in the face of assimilation and neo-colonial pressures (Lugo-Morin 2020; Grote et al. 2021).

5.14.1 Indigenous Food Systems

Indigenous peoples lead by far the most sustainable ways of living on the earth in its most fragile and marginal ecoregions. Often, their survival and that of the bioregions that they inhabit are intricately linked. It is through their extraordinary understanding of and relationship with their environment that they are able to ensure their subsistence. It is the continuity of these traditional practices which offers the global community options for sustainable futures. Indigenous knowledge enriches our understanding of the environment, adds value to the search for sustainability, and helps ensure the welfare of present and future generations. It is essential for the evolution of a new ethic that can assist the human race to live more sustainably. Traditional knowledge has accumulated over thousands of years, and it is held in the collective memory of societies. It is embodied in community practices, institutions, relationships, and rituals, in the annual rounds of work and ceremonies, and in the ethical teachings of elders. It is expressed through languages, myths, folklore, and the performance of arts such as music and dance. Knowledge is not only a record of the past, it is a resource for action in the present and the future. The essence of traditional knowledge is a deep understanding of the environment based on direct and close observation over a long period. This knowledge is an accumulative body. It is not usually a systematic body of knowledge articulated in logical progression, but it is closely integrated with the way of life. Knowledge of the environment is seen as an aspect of the knowledge of life. This knowledge is provisional and subject to testing; it is always open to change and must be constantly validated in the context of people's lives (Ulian et al. 2020; Ogunyiola et al. 2022).

5.14.2 Traditional Food Preservation Techniques

In times of abundance, food is processed and stored for times when food will be scarce. Traditional storage facilities found around the continent of Africa represent an important aspect of traditional knowledge in food security. Traditional storage structures consisted of clay pots, gourds, woven bags, wooden bins, granaries, and barns, underground pits, and rock caves. Up to 80 different types are used by the !Kung in Namibia and Botswana. These varying structures take into account the specific needs of a particular crop and the environment in which it must be stored to prevent insect infestation, disease development, germination, and spoilage. Usually, grains and legumes were stored in raised granaries, away from domestic animals with which humans share living space. With the introduction of metal tools and shift from agriculture as a dominant livelihood, an estimated 50 percent of traditional storage facilities have been abandoned. Metal tools rendered construction of traditional storage facilities too labor-intensive, and grain storage near fields has become popular as a means to quickly transport surplus grains to sell at market or store for cash crops. This has brought traditional storage systems to be replaced with metal bins or plastic bags, raising the question of how farmers might react to disease or insect infestation to crops stored in these containers, a useful practice when labor has already been committed to cultivating a different crop. A study in Zambia records that farmers may take action to protect the few crops stored in clay pots from those stored in modern containers by procuring and covering modern storage containers with the clay pots as a means to guarantee the safety of a small food supply should modern containers fail to prevent spoilage near an easily replaceable crop. This is an indication that traditional knowledge has not been forgotten, rather it is a resource waiting to be re-adapted to new circumstances (Okparavero et al. 2024; Kiran and Rampal 2024).

5.14.3 Role of Traditional Knowledge in Ensuring Food Availability

Higher crop diversity can also lead to improvements in diets as traditional crops are often more balanced and more nutritious than modern varieties. By maintaining traditional farming systems, crop varieties, and the associated knowledge, it is possible to maintain or increase crop diversity in many areas of the world (Imathiu 2021).

Ton That Thong found that marginalized communities in Vietnam who maintained shifting cultivation systems were doing so mainly to maintain crop diversity. When they settled and adopted modern agriculture, they began cultivating single varieties of crops and lost farm biodiversity. Commonly, this can be related to a loss of local knowledge on how to grow certain crops and the movement to modern varieties, which require less maintenance and are usually grown for higher yields. This has huge implications for food security should a crop fail due to a disease or change in climate (Quang and de Wit 2020; Kimkong et al. 2023).

In Orissa, tribal farmers in one district grew 92 different traditional rice varieties in the mid-1980s. 52 of these were lost after they began adopting Green Revolution rice cultivation, and 25 more disappeared because farmers were not paid for their annual maintenance of the seed. The remaining 15 varieties declined to seven or eight in number, and about five were being intensively cultivated. This illustrates a very clear case of how the adoption of modern farming methods can lead to loss of crop diversity (Ray 2022; Pathak et al. 2020).

Traditional knowledge can aid in maintaining crop diversity, which is a main concern in many small-scale farming systems where the loss of a crop variety can have serious repercussions. By using indigenous crop varieties and the associated farming knowledge, crop diversity can be maintained and sometimes increased. In the Americas, traditional farming systems have been shown to maintain 31% more crop species than modern systems in some areas (Kremen 2020; Sumberg and Giller 2022).

Traditional knowledge has a significant role to play in ensuring food access for future generations. It has the potential to complement scientific agriculture and increase yields at minimal cost to the environment. This is significant as changing climate patterns have led to a decrease in agricultural productivity in many areas of the world, including Africa (File and Derbile 2020; Kuyu and Bereka 2020).

5.14.4 Challenges and Threats to Traditional Knowledge

Traditional knowledge surrounding agriculture is rarely static and changes over time as new information and technologies become available. This, in itself, is not a problem and can even be a positive thing when new practices can be integrated with existing ones to increase yield or productivity. The problem comes when external influences cause a rapid decline in traditional knowledge, and that traditional knowledge then becomes lost or forgotten. There are various causes of external influence which can lead to a decline in traditional practices, including concentrated effort to increase productivity of land in a market-oriented system, external political pressures forcing change upon a community, or environmental disasters which render existing practices unviable. When this happens, traditional knowledge can be lost rapidly, as the incentive to continue with traditional practices diminishes, and a younger generation which has known nothing but the new practice grows up to replace the knowledge holders. The loss of traditional knowledge is most damaging when the new practice is less sustainable than the traditional practice, as it can lead to degradation of the environment and a decrease in food security for the community, which in turn can have negative impacts upon the livelihood and health of the people. This scenario leads to a cycle of decline in traditional knowledge as worsening livelihoods force people to seek further change to support themselves, at the expense of indigenous culture and knowledge. Measures to maintain traditional knowledge are therefore important to all communities, regardless of whether they are actively seeking to maintain knowledge or not. A worst-case scenario for the loss of traditional knowledge is that it cannot be revived or replaced, and the knowledge and the culture it represents are lost forever from the earth.

5.14.4.1 Loss of Traditional Practices

In addition to social change and globalization, another core driver in the decline of traditional agricultural knowledge is market integration. Opening up of markets, both nationally and internationally, have profound effects on wild and agricultural biodiversity throughout Africa. This is due to the fact that many products and biophysical resources gain new values once markets are accessible, and are often harvested to extinction. A common trend is the harvest of non-timber resources to supplement household income, as it is known that poorer households tend to adopt more unsustainable land use strategies. This was studied in the Zambian miombo, where an influx of male labour led to an increased hunting, which put pressure on wildlife populations. Simulation models have been used to predict that continued market integration will lead to the loss of 20–30% of wild resources in Africa, mainly in West, Central and Eastern regions.

Today, traditional agricultural practices are in decline throughout much of Africa. This is due in part to the fact that power in Africa is expressed through control of economic capital (salaried employment, cash cropping), as well as through political systems. These modern systems are far removed from traditional agricultural practices, where power was expressed through social capital, the disbursement of leadership, and systems of reciprocity and redistribution. This has led to loss of time and land intensive traditional systems to ecological approaches, which emulate modern agriculture. For example, it has been noted that during the colonial period in East Africa, terracing was widely practiced. It is now making a comeback due to its conservation capabilities. This trend, as well as the trend from traditional shifting cultivation to more intensified, sedentary forms of agriculture, effect a loss of traditional knowledge with each farmer generation. This is due to the complex and adaptive nature of much traditional agroecological knowledge, which is often only learned through trial and error.

5.14.4.2 Influence of Modern Agricultural Practices

The introduction of modern agricultural practices has dramatically affected traditional knowledge systems. Since colonial times, modern agricultural practices have displaced traditional systems aiming to increase production on the same piece of land. There has been a prioritization of annual crops over perennial and regenerating ones, mono-cropping over intercropping, and commercial crops over ones used for household consumption or barter. The modern trend has been towards simplification of ecosystems, with fewer crop varieties and higher dependence on purchased inputs, which increase production costs and exposure to risk. Modernization often decreases crop genetic diversity. It is estimated that 75 percent of crop genetic diversity was lost between 1900 and 2000. This is of major concern because crop genetic diversity is considered by many to be humanity's biological insurance policy. It provides the raw material for natural selection and adaptation to new biotic and abiotic stresses and changing environments. Traditional systems are highly dependent on local genetic diversity, wild crop relatives, and traditional farmer selection, all of which are threatened by the spread of modern varieties. The most influential modern agricultural input affecting traditional knowledge systems has been the adoption of Green Revolution technologies. The Green Revolution is a term used to describe the successful, yet controversial, development of technology transfer initiatives between the 1950s and the late 1960s. This involved the initial development and spread of high-yielding varieties of cereals, the dissemination of hybrid seeds, and the application of chemical inputs such as fertilizers and pesticides. The Green Revolution was largely initiated as a response to food security issues and the proliferation of the technologies was heavily promoted by international development agencies and donors. Impacts on traditional systems have been vast and often negative, while in some cases have also been highly beneficial.

5.14.5 Displacement of Indigenous Communities

Displacement of communities has become a common occurrence in Africa: due to political unrest, war and internal conflicts, and the setting up of conservation schemes and protected areas. The consequences of displacement are dire and can have serious effects on food security. Once displaced, people often lose contact with their pool of traditional knowledge as they are forcibly moved into an area where the environment and climate are different. This may mean the loss of an entire knowledge base as elders and other knowledge holders die before being able to pass it on to the next generation. Displacement may also lead to changed social structures and loss of community and collective approaches to farming and natural resources management. Typically, displaced people must resort to a more individual approach, often on smaller areas of land, and may be required to look for wage labor to survive. This usually means they have less time to devote to farming and maintenance of their land and resources. In the process, traditional knowledge is devalued and may be lost altogether.

5.14.6 Conservation and Revitalization of Traditional Knowledge

Integration of traditional and scientific knowledge recognizes that both knowledge systems have their strengths and weaknesses, and they are often complementary. It is also recognized that in many cases traditional knowledge has been marginalized or oppressed, and its validation will often require comparison with scientific knowledge or proof of concept. This should be a two-way process, where traditional knowledge is used to inform the scientific research and vice versa. In this way, the scientific knowledge can be used to enrich traditional knowledge for the benefit of the communities. An example of this has been traditional knowledge of the timing of seasonal indicators being used to inform climate scientists on the impacts of global warming, and the subsequent climate data being fed back to the communities to inform future agricultural decision-making.

Conservation and revitalization of traditional knowledge are crucial due to the current rapid erosion of indigenous knowledge systems as a result of the colonial era. It is widely recognized that traditional knowledge can be best utilized by its rightful owners within an environment that respects and facilitates their needs, so conservation efforts on traditional knowledge should be aimed at supporting indigenous communities in their goals. Documentation of traditional knowledge using a community-based approach is seen as a necessary first step in any conservation effort. This involves working with the knowledge holders in a way that helps to empower them, so as to ensure that any processes of documentation and future use of the documented knowledge are directed by the indigenous communities themselves. The documentation processes should be holistic, representing the complexity of traditional knowledge systems. This requires a flexible approach, recognizing that traditional knowledge is not static but rather an evolving body of knowledge. The most common tools of documentation include dictionaries and lexicons, life histories, and descriptions of customary laws and decision-making procedures as these hold the knowledge together.

5.14.6.1 Importance of Documenting Traditional Practices

Importance of documentations is stressed so as to make the knowledge readily available to the coming generations. Traditional agricultural practices are passed on from generation to generation. The trend of the youth of today moving to urban areas in search of "modern jobs" has resulted in break of this chain of knowledge. Documentations provide as a tool of learning for these youth as well as other community members who have not had the time or opportunity to learn the traditional practices. Instructional videos, leaflets or posters, or audio recordings can be used as means to deliver the information. The form of the documentation can vary depending on the culture and also the resources available. For example, marginalized communities in Kenya are using donkey-powered digital library in local languages [citation needed]. This library contains a wide range of materials including oral literature, agricultural practices, health, and many other things. The creation of the materials involved several training workshops for community members led by professionals on the use of digital recording equipment, computer skills, and video making. In this case, the knowledge on how to create these materials is just as important as the material itself.

5.14.7 Integration of Traditional and Scientific Knowledge

The research and development process should begin with the identification of the needs of the resource user. These should be ranked in terms of their importance to the resource users and not to the funding agency or researcher, as the case often currently seems to be. Once the needs are identified, research should begin to evaluate the existing knowledge, practices, and technologies that pertain to the needs. At this stage, one should utilize the fact that different knowledge systems often deal with the same issues, but in different ways. A true understanding of the resource needs will allow the scientific researcher to evaluate and fully understand the traditional methods. It may be that the problems being addressed have been created or exacerbated by outside influences. In these cases, it may be that a reversion to a traditional practice adapted with new knowledge may be the best solution. On the other hand, there may be a need to integrate new knowledge and technologies. In this situation, it will be possible to identify the factors that determine the success or failure of a given practice or technology, thus providing a framework for its future testing and adaptation.

When traditional knowledge has been scientifically tested and subsequently adapted to local conditions, and when local conditions have been taken into account in the development of new practices, the results have often been quite successful. However, the history of such activities is fraught with examples of the utilization of such knowledge, often with no benefit to the holders of the knowledge. To be successful in agricultural endeavors that target marginal areas and the world's poor, a new research and development paradigm is needed. This suggests that the process of interaction between traditional and scientific knowledge systems should be reoriented.

5.14.8 Empowering Indigenous Communities

One approach to helping indigenous people is to improve their livelihoods and control over development in ways that are consistent with their own aspirations and needs. This may entail the enhancement of local-level policy making and economic and social activities that are entrepreneurial but based on the maintenance of their own cultural identity and natural resource management systems. To have meaningful informed decision making and full participation, it is important for indigenous communities not to be heavily weighted with the negative aspects of development activities. It is important to find ways and means for them to capitalize on the competencies they already have and to maintain and further develop their traditional knowledge systems.

Indigenous people and rural communities derive their identities from their ancestral territories. Their livelihoods and resilience against external forces have maintained the vitality of their cultures. Today, however, indigenous peoples and their communities are subject to new external forces that may have a negative impact on their traditional knowledge and livelihoods. Globalization has made the world a smaller place and has increased interactions among societies. More recent market liberalization, resource privatization, and state decentralization through deregulation and "austerity measures" are forces to which indigenous communities in upland and remote areas are increasingly subjected. All too often, these forces are detrimental to indigenous livelihoods and environments. Taken as a whole, they have resulted in a loss of control over traditional territories, degradation of resources, and destruction of indigenous sustainable development systems.

5.14.9 Policy and Institutional Support for Traditional Knowledge

Specific measures for protection of traditional knowledge in national policies have included the development of sui generis legislation, which has occurred in countries such as Fiji and Papua New Guinea, and more recently the development of the Andean community legislation. Other measures have included the formation of a national policy on bioprospecting and traditional knowledge research, and the development of a code of ethics for researchers working with traditional knowledge. The most common approach to date has been the formation of an ad-hoc committee or unit for traditional knowledge within a relevant government department.

There are several areas that are relevant to policy development surrounding the legal status and ownership of traditional knowledge. These include trade, environment, health, agriculture, education, and the need for international law or agreements. Traditional knowledge has been addressed in several national policies in different ways, depending on the importance of traditional knowledge to the nation and the awareness about traditional knowledge and its role in development. In countries such as Fiji and some West African countries, traditional knowledge is highly integrated into national policy, whereas in many other countries, policy for traditional knowledge may be confined to international negotiations and a few areas such as environment and health.

5.14.9.1 Recognition of Traditional Knowledge in National Policies

There has been no shortage of policy statements attesting to the importance of traditional knowledge in agriculture throughout Africa. However, policy reforms have been implemented in only a few countries, and assessment of their impact is difficult to measure. Three countries discussed here provide examples of different methods to recognize traditional knowledge in agricultural policy. These include Tanzania, which has made significant strides by producing a policy specifically for the conservation of crop genetic resources and strengthening ties with the formal sector in plant breeding and farmer participation. Rice is highlighted as a target crop for this policy, and research in the Morogoro region is used as a case study to assess its effectiveness. Mention is also made of Kenya, a country which has demonstrated a long-standing commitment to including traditional farming systems within national agricultural research strategies. Steps taken in Mali have been much more recent; however, AKDN's involvement in the IFAD funded Programme for the Promotion of Market Gardening in Bamako has demonstrated how an innovative intervention between traditional and market-oriented farming systems can be reached to benefit smallholder communities. In Mali, a recent policy shift has encouraged participation in decentralized rural development programs through the process of transferring knowledge and skills between communities, and these have proven to be an effective venue for the interaction and integration of traditional knowledge in agricultural innovations.

African governments have begun to recognize the failure of past agricultural and food security policies and have initiated policy reforms. Based on the acknowledgement of the economic, ecological, and social importance of smallholder agriculture, some national policies have incorporated the need to sustain and strengthen traditional farming systems. The integration of traditional knowledge into agricultural policy is a key step in achieving food security and is the focus of this chapter.

5.14.9.2 Collaboration with Indigenous Organizations

In the same way, tools could be developed directly with the users to map the traditional knowledge relevant to a certain piece of genetic material or a new agricultural technique being introduced. This process would increase awareness within the indigenous community, improving communication between them and outside users of TK on how it can be best protected and to avoid any uncompensated usage. The employment of such information technologies is directly relevant to a type of policy already exists but with little success - custodian programs where government or private institutions employ indigenous people to be knowledge custodians in order to advise which types of research or development the community will permit. This is one area where TK has global importance as a tool for sustainable development and poverty eradication, but only if the knowledge is not exploited.

Rheinhard Leimeister, Helmut Krcmar, Christine Veldhoen, and Stefan Fischer discuss their design for virtual communities for indigenous knowledge management. They see virtual communities as an organizing structure for social interaction and exchange of information among people who share tacit knowledge and want to create a holistic model of the knowledge processes within virtual communities relevant to indigenous people. This could be seen as an elaborate and more technological approach to making Article 8j effective. This kind of methodology would empower indigenous communities in negotiating the terms and conditions under which outside access to their knowledge may be permitted.

This section was written by Nadine Cronin, who is the Director of the Cross Cultural Awareness Programme and Co-Chair of Te Taumata Kaumatua—Maori Elders Council Health. In it, she discusses strategies for collaboration based on New Zealand experiences. She concludes with "collaboration with indigenous organisations in the development and implementation of policy related to TK facilitates a process for communities to engage with governments and research institutions in defining the terms and conditions under which outside access and use of their knowledge will be permitted and the sharing of the benefits." Cronin's analysis is a simplified version of the CBD's vision written into Article 8j.

5.14.9.3 Funding and Resources for Traditional Knowledge Initiatives

External funding sources, whether international development assistance or philanthropic donations, may be quite volatile. Therefore, it is important to develop sustainable sources of funding for traditional knowledge research. One option is to link traditional knowledge initiatives to private sector or community-based initiatives, and in some cases this might work. However, it is likely that in the medium-term future, agricultural research needs long-term investment. One potential source of stable funding is national-level contributions from African countries themselves. This could be derived from general tax revenues or specific levies on natural resource use in the case of traditional knowledge associated with natural resource management. However, there will need to be a negotiated balance between national priorities and externally driven research agendas.

Funding of traditional knowledge initiatives does remain a great challenge. There is severe underfunding for agricultural research in general, and traditional knowledge initiatives specifically, in Africa. Governmental support for traditional knowledge research has been inadequate, and there is little evidence of an improvement in this area in recent years. International organisations have provided some financial support for traditional knowledge initiatives in Africa, and this has become an increasing trend over the past decade. The co-sponsorship program between the International Development Research Centre and the Canadian International Development Agency is one noteworthy example of a funding initiative for traditional knowledge research. However, in recent years official development assistance to agriculture has been declining. There are concerns that in the current global economic climate it will be difficult to maintain and increase the levels of funding for traditional knowledge research.

5.14.10 Case Studies

The case studies were primary data sources obtained from field research conducted between 1997 and 2001 in Tanzania, Zambia, and Botswana. Mixed qualitative and quantitative research methods were employed, including structured interviews, participant observation, and informal discussion. Traditional leaders, extension workers, farmers, and women's groups were interviewed using structured questionnaires. Sampling of respondents for interviews was purposive in order to obtain the views of key informants with various experiences of agriculture and livelihoods. Random sampling and snowball sampling were then employed to obtain a wider range of views and experiences. Data from questionnaire surveys were supplemented with information from participant observation and informal discussion primarily in order to gain a deeper understanding of the local context.

This section contains case studies from different regions of Africa centered on the themes of successful implementation of traditional knowledge in agriculture and the impact of traditional knowledge on food security. The case studies have been selected to provide empirical evidence of the issues discussed in previous sections and to illuminate the dynamic nature of traditional knowledge and its potential in aiding the development of small-scale agriculture and food security in local communities.

5.14.10.1 Successful Implementation of Traditional Knowledge in Agriculture

The success of traditional knowledge in many areas of agriculture has led to several projects aimed at promoting and improving traditional methods. One such project was conducted in southern Mali and Burkina Faso where farmer experimentation with traditional soil and water conservation practices was fostered through the formation of farmer-managed research groups. In areas where marginalization has led to loss of confidence in traditional knowledge, participatory projects have been initiated to revitalize traditional methods as in the case of the aforementioned organic pest control methods in Uganda. National policy has a large impact on the success of traditional agriculture and where policies are poorly informed about the benefits of traditional knowledge, sustainability and food security become jeopardized. This was seen in Malawi when the government traded a 10-year debt relief package with the World Bank and IMF for structural adjustment programs imposed by the Washington Consensus. The eventual cutting of subsidies to maize producers led to the increased use of hybrid maize and chemical fertilizers and the neglect of traditional maize varieties. This had a negative impact on food security and biodiversity in Malawi (Tiwari 2022; Di et al. 2021).

This is highly evident in Africa, where traditional knowledge is key in the success of many small-scale farmers. Traditional knowledge is used in many facets of agriculture including crop selection, seed saving, soil fertility management, plant propagation, and crop protection. In a study conducted by the Centre for Development of Best Practices in Health (CDBPH) in Uganda, several traditional methods of organic pest control were tested and found to be as effective as synthetic chemicals. These methods were low cost and utilized locally available resources and thus more sustainable than synthetic pesticides. Traditional methods of conservation such as seed saving and crop variety selection are employed by thousands of small-scale farmers. One example is the seed saving of traditional varieties of staple crops such as maize, millet, rice, and beans. These varieties are often more nutritious than introduced varieties and are highly adapted to local environmental conditions.

In a series of discussions with the Ahanta community, it became clear that they did not want an external injection of funds in the form of a food aid program or the introduction of an alternative income-generating activity. They rather wanted to increase food production in the hope of reducing food prices and increasing the purchasing power of the community (Dzanku et al. 2023; Takyiakwaa et al. 2024; Ameyaw et al. 2020; Takyiakwaa et al. 2021; Ovadia et al. 2020).

A major constraint to food production was the lack of appropriate tools that had resulted from the decline in traditional apprenticeship in woodwork. This is because the apprentice system broke down as young people sought the quick money that could

Fig. 5.1 Monoculture plantation of banana (Photo by: E.D. Wiafe)

be earned in the local fishing industry. Fishing had become three to four times more profitable as a result of increased demand from other communities that were also experiencing food shortages. This had created a vicious circle as the fishing income enabled people to buy food, thus reducing the time available to farm and increasing the dependence on purchasing food from other communities. The consequence was higher fish prices and further increased fishing activity (Palanivel and Shah 2021).

The Ahanta community in the western region of Ghana experiences regular seasonal food shortages. These traditionally occurred during the rainy season when food from the previous harvest began to run out. The community usually coped with food shortages by accumulating debts to be repaid after the next harvest or by selling family assets. However, ActionAid worked with the community to address their food security problem (Fig. 5.1). PRA identified the food types families rely on between harvests and the coping strategies used to obtain food during the hunger season. This information was then used to identify the constraints to food production and consumption (Mohammed et al. 2023; Asare and Ernest 2021; Dzanku et al. 2023; Fordjour 2023).

References

Abbas M, Saleem M, Hussain D, Ramzan M, Jawad Saleem M, Abbas S et al (2022) Review on integrated disease and pest management of field crops. Int J Tropical Insect Sci 42(5):3235–3243. https://doi.org/10.1007/s42690-022-00753-4

Abera G, Gerkabo H (2021) Effects of green manure legumes and their termination time on yield of maize and soil chemical properties. Arch Agron Soil Sci 67(12):1657–1671. https://doi.org/10.1080/03650340.2020.1818346

Adetunji AT, Ncube B, Mulidzi R, Lewu FB (2020) Management impact and benefit of cover crops on soil quality: a review. Soil Tillage Res 204:104717. https://doi.org/10.1016/j.still.2020.104717

Adetutu MO, Ajayi V (2020) The impact of domestic and foreign R&D on agricultural productivity in sub-Saharan Africa. World Dev 125:104690. https://doi.org/10.1016/j.worlddev.2019.104690

Akinola R, Pereira LM, Mabhaudhi T, De Bruin FM, Rusch L (2020) A review of indigenous food crops in Africa and the implications for more sustainable and healthy food systems. Sustainability 12(8):3493. https://doi.org/10.3390/su12083493

Amede T, Legesse G, Agegnehu G, Gashaw T, Degefu T, Desta G et al (2021) Short term fallow and partitioning effects of green manures on wheat systems in East African highlands. Field Crop Res 269:108175. https://doi.org/10.1016/j.fcr.2021.108175

Ameyaw AB, Breckwoldt A, Reuter H, Aheto DW (2020) From fish to cash: analyzing the role of women in fisheries in the western region of Ghana. Marine Policy 113:103790. https://doi.org/10.1016/j.marpol.2019.103790

Asare AA, Ernest KA (2021) Geotechnical and geochemical assessment of natural sands in Ahanta West Area, Southwestern Ghana. J Geosci Environ Protect 9(2):75–92. https://doi.org/10.4236/gep.2021.92006

Baker BP, Green TA, Loker AJ (2020) Biological control and integrated pest management in organic and conventional systems. Biol Control 140:104095. https://doi.org/10.1016/j.biocontrol.2019.104095

Barakagira A, Ndungo I (2023) Watershed management and climate change adaptation mechanisms used by people living in dryland areas of Lokere catchment in Karamoja, Uganda. Environ Socio-Econ Stud 11(2):45–56. https://doi.org/10.2478/environ-2023-0012

Barrios E, Gemmill-Herren B, Bicksler A, Siliprandi E, Brathwaite R, Moller S et al (2020) The 10 Elements of Agroecology: enabling transitions towards sustainable agriculture and food systems through visual narratives. Ecosyst People 16(1):230–247. https://doi.org/10.1080/26395916.2020.1808708

Britwum K, Demont M (2022) Food security and the cultural heritage missing link. Glob Food Secur 33:100620. https://doi.org/10.1016/j.gfs.2022.100620

Bunch R (2023) Green manure/cover crops for recuperating soils and maintaining soil system fertility. In: Biological approaches to regenerative soil systems. Academic Press, pp 59–86. https://doi.org/10.1016/B978-012-821405-9.00004-3

Campi M, Dueñas M, Fagiolo G (2021) Specialization in food production affects global food security and food systems sustainability. World Dev 141:105411. https://doi.org/10.1016/j.worlddev.2020.105411

Cardinael R, Guibert H, Brédoumy STK, Gigou J, N'Goran KE, Corbeels M (2022) Sustaining maize yields and soil carbon following land clearing in the forest–savannah transition zone of West Africa: Results from a 20-year experiment. Field Crop Res 275:108335. https://doi.org/10.1016/j.fcr.2021.108335

Clapp J, Moseley WG, Burlingame B, Termine P (2022) The case for a six-dimensional food security framework. Food Policy. 106:102164. https://doi.org/10.1016/j.foodpol.2021.102164

Crystal-Ornelas R, Thapa R, Tully KL (2021) Soil organic carbon is affected by organic amendments, conservation tillage, and cover cropping in organic farming systems: a meta-analysis. Agric Ecosyst Environ 312:107356. https://doi.org/10.1016/j.agee.2021.107356

Daum T, Birner R (2020) Agricultural mechanization in Africa: myths, realities and an emerging research agenda. Global food security. 26:100393. https://doi.org/10.1016/j.gfs.2020.100393

De Waal A (2020) Governance implications of epidemic disease in Africa: updating the agenda for COVID-19. African Affairs 119(477):548–560. https://doi.org/10.1093/afraf/adaa019

Deka N, Goswami K (2021) Economic sustainability of organic cultivation of Assam tea produced by small-scale growers. Sustain Prod Consum 27:1111–1124. https://doi.org/10.1016/j.spc.2021.02.016

Di Sacco A, Hardwick KA, Blakesley D, Brancalion PHS, Breman E, Cecilio Rebola L et al (2021) Ten golden rules for reforestation to optimize carbon sequestration, biodiversity recovery and livelihood benefits. Glob Change Biol 27(7):1328–1348. https://doi.org/10.1111/gcb.15498

Dickens C, Whitney C, Luedeling E, Dlamini V, O'Brien G, Greffiths IJ (2023) Environmental flows in support of sustainable intensification of agriculture in the Letaba River Basin, South Africa. IWMI Research Report. International Water Management Institute. https://www.iwmi.cgiar.org

Diop M, Chirinda N, Beniaich A, El Gharous M, El Mejahed K (2022) Soil and water conservation in Africa: State of play and potential role in tackling soil degradation and building soil health in agricultural lands. Sustainability 14(20):13425. https://doi.org/10.3390/su142013425

Dossou JAD, Diogo RVC, Gbedjisokpa MG, Seidou AB, Awéha GA, Paul BK (2022) Agronomic evaluation of the effects of two green manure cover crops on maize (*Zea mays*) cultivation in four-agroecological zones of Benin. In: Proceedings of the Australian Agronomy Conference. agronomyaustraliaproceedings.org

Dzanku FM, Asante KT, Hodey LS (2023) Heterogeneous market participation channels and household welfare. Oxford Dev Stud 51(3):317–334. https://doi.org/10.1080/13600818.2022.2105457

Emeana EM, Trenchard L, Dehnen-Schmutz K (2020) The revolution of mobile phone-enabled services for agricultural development (m-Agri services) in Africa: the challenges for sustainability. Sustainability 12(2):485. https://doi.org/10.3390/su12020485

Esfandiar K, Dowling R, Pearce J, Goh E (2020) Personal norms and the adoption of pro-environmental binning behaviour in national parks: an integrated structural model approach. J Sustain Tour 28(1):10–32. https://doi.org/10.1080/09669582.2019.1657539

Falayi M, Gambiza J, Schoon M (2021) A scoping review of environmental governance challenges in southern Africa (2010–2020). Environ Conserv 48(4):305–315. https://doi.org/10.1017/S0376892921000246

Fastovich GG, Kapsargina SA (2020) Institute of import substitution in the agro-industrial complex of modern Russia (theoretical and legal approach). IOP Conf

Ser Earth Environ Sci 421(3):032035. https://doi.org/10.1088/1755-1315/421/3/032035

File DJ, Derbile EK (2020) Sunshine, temperature and wind: community risk assessment of climate change, indigenous knowledge and adaptation planning in Ghana. Int J Clim Change Strateg Manag 12(1):22–38. https://doi.org/10.1108/IJCCSM-10-2018-0070

Fordjour R (2023) Cultural tourism and socio-economic recovery of an ex-urban community: the case of Ankos masquerade festival (Effia Kuma), Takoradi western region of Ghana. J Tour Cult Change

Gashure S, Wana D, Samimi C (2022) Impacts of climate variability and climate-smart agricultural practices on crop production in UNESCO designated cultural landscapes of Konso, Ethiopia. Theor Appl Climatol 150(3):1495–1511. PMC

Giller KE, Delaune T, Silva JV, Descheemaeker K, van de Ven G, Schut AG et al (2021) The future of farming: Who will produce our food? Food Secur 13(5):1073–1099. https://doi.org/10.1007/s12571-021-01184-6

Giller KE, Delaune T, Silva JV, van Wijk M, Hammond J, Descheemaeker K et al (2021) Small farms and development in sub-Saharan Africa: Farming for food, for income or for lack of better options? Food Secur 13(6):1431–1454. https://doi.org/10.1007/s12571-021-01209-0

Grebremedhin B, Hoekstra D, Tegegne A (2006) Commercialization of Ethiopian agriculture: extension service from input supplier to knowledge broker and facilitator. International Livestock Research Institute (ILRI). https://hdl.handle.net/10568/1917

Grote U, Fasse A, Nguyen TT, Erenstein O (2021) Food security and the dynamics of wheat and maize value chains in Africa and Asia. Front Sustain Food Syst 4:617009. https://doi.org/10.3389/fsufs.2020.617009

Harwood RR (2020) A history of sustainable agriculture. In: Francis CA (ed) Sustainable agricultural systems. CRC Press, pp 3–19. https://doi.org/10.1201/9780429429853

Hill MP, Conlong D, Zachariades C, Coetzee JA, Paterson ID, Miller BE et al (2021) The role of mass-rearing in weed biological control projects in South Africa. Afr Entomol 29(3):1030–1044. https://doi.org/10.4001/003.029.1030

Hossain A, Krupnik TJ, Timsina J, Mahboob MG, Chaki AK, Farooq M et al (2020) Agricultural land degradation: processes and problems undermining future food security. In: Hasanuzzaman M (ed) Environment, climate, plant and vegetation growth. Springer, Cham, pp 17–61. https://doi.org/10.1007/978-3-03049732-3_2

Hussain MI, Farooq M, Muscolo A, Rehman A (2020) Crop diversification and saline water irrigation as potential strategies to save freshwater resources and reclaim marginal soils: a review. Env Sci Pollut Res 27(23):28695–28729. https://doi.org/10.1007/s11356-020-09111-8

Hussain S, Hussain S, Guo R, Sarwar M, Ren X, Krstić D et al (2021) Carbon sequestration to avoid soil degradation: a review on the role of conservation tillage. Plants 10(10):2001. https://doi.org/10.3390/plants10102001

Ibrahim ID, Hamam Y, Alayli Y, Jamiru T, Sadiku ER, Kupolati WK et al (2021) A review on Africa energy supply through renewable energy production: Nigeria, Cameroon, Ghana and South Africa as a case study. Energy Strategy Rev 38:100740. https://doi.org/10.1016/j.esr.2021.100740

Imathiu S (2021) Neglected and underutilized cultivated crops with respect to indigenous African leafy vegetables for food and nutrition security. J Food Secur 9(3):115–125

Irawan A (2023) The smallholder coffee farmer's livelihood adaptation strategies in Bengkulu, Indonesia. J Strategy Manag 16(2):287–303. https://doi.org/10.1108/JSMA-06-2022-0121

Kansanga MM, Kerr RB, Lupafya E, Dakishoni L, Luginaah I (2021) Does participatory farmer-to-farmer training improve the adoption of sustainable land management practices? Land Use Policy 108:105477. https://doi.org/10.1016/j.landusepol.2021.105477

Karanja Ng'ang'a S, Jalang'o DA, Girvetz EH (2020) Adoption of technologies that enhance soil carbon sequestration in East Africa: What influence farmers' decision? Int soil water Conserv Res 8(1):90–101. https://doi.org/10.1016/j.iswcr.2020.01.001

Kato E, Mekonnen DK, Ringler C (2022) Gender gaps in sustainable land management and implications for agricultural productivity: evidence from Ethiopia. World Dev 155:105872. https://doi.org/10.1016/j.worlddev.2022.105872

Khoury CK, Brush S, Costich DE, Curry HA, De Haan S, Engels JM et al (2022) Crop genetic erosion: understanding and responding to loss of crop diversity. N Phytologist 233(1):84–118. https://doi.org/10.1111/nph.17733

Kim N, Zabaloy MC, Guan K, Villamil MB (2020) Do cover crops benefit soil microbiome? A meta-analysis of current research. Soil Biol Biochem 142:107701. https://doi.org/10.1016/j.soilbio.2019.107701

Kimkong H, Promphakping B, Hudson H, Day SCJ (2023) Agricultural transformation in the rural farmer communities of Stung Chrey Bak, Kampong Chhnang Province, Cambodia. Agriculture 13(2):389. https://doi.org/10.3390/agriculture13020389

Kiran Babu P, Rampal (2024) Traditional ecological knowledge in sustainable conservation of seeds and food grains in the Himalayas. In: Addressing the climate crisis in the Indian Himalayas: can traditional ecological knowledge help? Springer Nature, Cham, pp 53–93. https://doi.org/10.1007/978-3-031-46042-5_3

Koudahe K, Allen SC, Djaman K (2022) Critical review of the impact of cover crops on soil properties. Int Soil Water Conserv Res 10(3):343–354. https://doi.org/10.1016/j.iswcr.2022.01.002

Kremen C (2020) Ecological intensification and diversification approaches to maintain biodiversity, ecosystem services and food production in a changing world. Emerg Top Life Sci 4(3):229–240. https://doi.org/10.1042/ETLS20190123

Kuyu CG, Bereka TY (2020) Review on contribution of indigenous food preparation and preservation techniques to attainment of food security in Ethiopian. Food Sci Nutr 8(1):3–15. https://doi.org/10.1002/fsn3.1274

Labeyrie V, Renard D, Aumeeruddy-Thomas Y, Benyei P, Caillon S, Calvet-Mir L et al (2021) The role of crop diversity in climate change adaptation: insights from local observations to inform decision making in agriculture. Curr Opin Environ Sustain 51:15–23. https://doi.org/10.1016/j.cosust.2021.01.005

Lal R (2020) Soil organic matter and water retention. Agronomy J 112(5):3265–3277. https://doi.org/10.1002/agj2.20282

Lieber M, Chin-Hong P, Whittle HJ, Hogg R, Weiser SD (2021) The synergistic relationship between climate change and the HIV/AIDS epidemic: a conceptual framework. AIDS Behav 25:2266–2277. https://doi.org/10.1007/s10461-020-03080-0

Lindsjö K, Mulwafu W, Andersson Djurfeldt A, Joshua MK (2021) Generational dynamics of agricultural intensification in Malawi: Challenges for the youth and elderly smallholder farmers. Int J Agric Sustain 19(5–6):423–436. https://doi.org/10.1080/14735903.2021.1882393

Lugo-Morín DR (2020) Indigenous communities and their food systems: a contribution to the current debate. J Ethn Foods 7. Article 34. MDPI

Mohammed AS, Tuokuu FXD, Adda EB (2023) Livelihood access and challenges of coastal communities: insights from Ghana. J Glob Responsibility 14(4):452–475. https://doi.org/10.1108/JGR-01-2023-0005

Mpumi N, Machunda RS, Mtei KM, & Ndakidemi PA (2020) Selected insect pests of economic importance to Brassica oleracea, their control strategies and the potential threat to environmental pollution in Africa. Sustainability 12(10):4034. https://doi.org/10.3390/su12104034

Nafi E, Webber H, Danso I, Naab JB, Frei M, Gaiser T (2020) Interactive effects of conservation tillage, residue management, and nitrogen fertilizer application on soil properties under maize-cotton rotation system on highly weathered soils of West Africa. Soil Tillage Res 196:104473. https://doi.org/10.1016/j.still.2019.104473

Nin-Pratt A (2021) Agricultural R&D investment intensity: a misleading conventional measure and a new intensity index. Agric Econ 52(3):327–342. https://doi.org/10.1111/agec.12623

Ogunyiola A, Gardezi M, Vij S (2022) Smallholder farmers' engagement with climate-smart agriculture in Africa: The role of local knowledge and upscaling. Climate Policy 22(6):725–739. https://doi.org/10.1080/14693062.2022.2030914

Okparavero NF, Grace OO, Rukayat Q, Jimoh OM, Ishola TD, Okunlade AF et al (2024) Effective Storage Structures for Preservation of Stored Grains in Nigeria: a Review. Ceylon J Sci 53(1):139–147. https://doi.org/10.4038/cjs.v53i1.8146

Olufolaji DB, Ajayi AM (2021) Eco-friendly management of plant diseases in sub-Saharan Africa. Indian Phytopathol 74(2):347–359. https://doi.org/10.1007/s42360-021-00352-9

Oluma A, Abadiga M, Mosisa G, Etafa W, Fekadu G (2020) Food insecurity among people living with HIV/AIDS on antiretroviral therapy at public hospitals of Western Ethiopia. Int J Food Sci 2020. Article 9026012. https://doi.org/10.1155/2020/9026012

Ovadia JS, Ayelazuno JA, Van Alstine J (2020) Ghana's petroleum industry: expectations, frustrations and anger in coastal communities. J Mod Afr Stud 58(3):397–424. https://doi.org/10.1017/S0022278X20000219

Palanivel H, Shah S (2021) Unlocking the inherent potential of plant genetic resources: food security and climate adaptation strategy in Fiji and the Pacific. Environ Dev Sustain 23(6):8573–8592. https://doi.org/10.1007/s10668-020-00942-1

Pathak H, Tripathi R, Jambhulkar NN, Bisen JP, Panda BB (2020) Eco-regional-based rice farming for enhancing productivity, profitability and sustainability. Indian Council of Agricultural Research (ICAR). https://icar.gov.in

Pawlak K, Kołodziejczak M (2020) The role of agriculture in ensuring food security in developing countries: considerations in the context of the problem of sustainable food production. Sustainability 12(13):5488. https://doi.org/10.3390/su12135488

Piemontese L, Castelli G, Fetzer I, Barron J, Liniger H, Harari N et al (2020) Estimating the global potential of water harvesting from successful case studies. Glob Env change 63:102121. https://doi.org/10.1016/j.gloenvcha.2020.102121

Quang NM, de Wit J (2020) Transformative learning and grassroots climate adaptation: case studies in Vietnam's Mekong Delta. Nat Conserv Res 5(Suppl 1):34–45

Ratemo CM, Ogendi GM, Huang G, Ondieki RN (2020) Application of traditional ecological knowledge in food and water security in the Semi-Arid Turkana County, Kenya. Open J Ecol 10(6):341–355. https://doi.org/10.4236/oje.2020.106021

Ratto F, Bruce T, Chipabika G, Mwamakamba S, Mkandawire R, Khan Z et al (2022) Biological control interventions and botanical pesticides for insect pests of crops in sub-Saharan Africa: a mapping review. Front Sustain Food Sys 6:827164. https://doi.org/10.3389/fsufs.2022.827164

Ravisankar N, Ansari MA, Panwar AS, Aulakh CS, Sharma SK, Suganthy M et al (2021) Organic farming research in India: potential technologies and way forward. ICAR–Indian Institute of Farming Systems Research. icar.gov.in

Ray A (2022) Disappearance of autumn (aus) rice, entry of HYVs and environmental sustainability in a Green Revolution state of eastern India. World Dev Sustain 1:100018. https://doi.org/10.1016/j.wds.2022.100018

Sango CO (2021) Factors affecting public housing development projects in Nairobi County. Int J Innov Res Dev 10(4):1–11

Sawicka B, Egbuna C (2020) Pests of agricultural crops and control measures. In: Egbuna C, Gaiya SAA (eds) Natural remedies for pest, disease and weed control. Academic Press, pp 1–16. https://doi.org/10.1016/B978-0-12-819304-4.00001-5

Shahzad R, Jamil S, Ahmad S, Nisar A, Amina Z, Saleem S et al (2021) Harnessing plant transcription factors to develop climate-resilient crops for global food security. Saudi J Biol Sci 28(4):2323–2341. https://doi.org/10.1016/j.sjbs.2021.01.038

Sullivan CA, Meigh JR, Giacomello AM (2003) The water poverty index: development and application at the community scale. Nat Resour Forum 27(3):189–199. https://doi.org/10.1111/1477-8947.00054. IDEAS/RePEc

Sumberg J, Giller KE (2022) What is 'conventional' agriculture? Global Food Secur 32:100616. https://doi.org/10.1016/j.gfs.2022.100616

Takyiakwaa D, Tetteh PSK, Asante KT (2021) Explaining the weakness of associational life in oil palm growing communities in southwestern Ghana. APRA working paper 61. https://doi.org/10.19088/APRA.2021.061

Takyiakwaa D, Tetteh PSK, Asante KT (2024) 'Associations do not survive here': inequality, mistrust and obstacles to collective action in oil palm growing communities in Ghana. J Asian Afr Stud 59(3):482–498. https://doi.org/10.1177/00219096231224684

Thornton P, Dijkman J, Herrero M, Szilagyi L, Cramer L (2022) Aligning vision and reality in publicly funded agricultural research for development: a CGIAR case study. Food Policy 109:102253. https://doi.org/10.1016/j.foodpol.2022.102253

Tiwari SP (2022) Information and communication technology initiatives for knowledge sharing in agriculture. arXiv Preprint, arXiv:2202.08649. https://arxiv.org/abs/2202.08649

Uebersax MA, Cichy KA, Gomez FE, Porch TG, Heitholt J, Osorno JM et al (2023) Dry beans (*Phaseolus vulgaris* L.) as a vital component of sustainable agriculture and food security. Legume Sci 5(1):e155. https://doi.org/10.1002/leg3.155

Ulian T, Diazgranados M, Pironon S, Padulosi S, Liu U, Davies L et al (2020) Unlocking plant resources to support food security and promote sustainable agriculture. Plants People Planet 2(5):421–445. https://doi.org/10.1002/ppp3.10145

Viana CM, Freire D, Abrantes P, Rocha J, Pereira P (2022) Agricultural land systems importance for supporting food security and sustainable development goals: a systematic review. Sci Total Env 806:150718. https://doi.org/10.1016/j.scitotenv.2021.150718

Wezel A, Herren BG, Kerr RB, Barrios E, Gonçalves ALR, Sinclair F (2020) Agroecological principles and elements and their implications for transitioning to sustainable food systems. A review. Agron Sustain Dev 40:1–13. https://doi.org/10.1007/s13593-020-00646-z

Yeboah FK, Jayne TS (2020) Africa's evolving employment trends. In Jayne TS et al (eds) The transformation of rural Africa. Michigan State University Press, pp 39–62

Yi X, Sauer J, Yin C, Li F (2022) Explaining farmers' reluctance to adopt green manure cover crops in Northwest China. J Integr Agric 21(4):1056–1067. https://doi.org/10.1016/S2095-3119(21)63632-9

Żmija K, Fortes A, Tia MN, Šūmane S, Ayambila SN, Żmija D et al (2020) Small farming and generational renewal in the context of food security challenges. Glob Food Secur 26:100412. https://doi.org/10.1016/j.gfs.2020.100412

Zsögön A, Peres LE, Xiao Y, Yan J, Fernie AR (2022) Enhancing crop diversity for food security in the face of climate uncertainty. Plant J 109(2):402–414. https://doi.org/10.1111/tpj.15504

Wildlife Conservation Initiatives in Africa 6

6.1 Overview of Wildlife Conservation Initiatives in Africa

The African continent is home to the largest variety of wildlife on the planet. As the world's second-largest and second-most-populous continent, with an estimated population of 1.2 billion, Africa's human population is increasing at an alarming rate. The demand for land for agricultural use, settlements, irrigation, and land development continues to rise in order to support this population. As a result, tsetse flies, the vectors for sleeping sickness in humans and nagana in livestock, have been an important target for eradication schemes in South and North Africa. By clearing patches of land for cultivation, these have, in the past, inadvertently led to the spread of invasive species into new territories. Pest species such as the tsetse fly, rats, and certain species of snails can cause substantial damage to crops and livestock and are a threat to human health. The alteration of land has also resulted in the decline of certain wildlife species as it has led to fragmentation and destruction of their habitats. This is a major factor in the continued trend of decline in both the populations and distribution of many species. Having large rural populations and often widespread poverty, political and social frameworks in many African nations have placed insufficient importance on environmental management and the sustainable and equitable use of natural resources. Corruption in politics has also led to the exploitation of resources and has contributed to the declining management of wilderness areas. These factors have contributed to the exploitation of resources such as minerals, timber, and wildlife. Unsustainable exploitation of wildlife through commercial hunting, poaching, and trade in wildlife products has resulted in the loss of many animal species. In many instances, this has threatened the existence of viable populations of various species, and in some cases, extinction has been documented. For example, the West African black rhinoceros was declared extinct by the IUCN in 2011. Due to the continued exploitation of these resources and the lack of benefit received by the local communities, these conservation-reliant species continue to be threatened.

6.2 Importance of Wildlife Conservation

Awakening to the significant loss of biodiversity, the world community has now made a partial commitment to deal with the problems associated with natural resource degradation and biological diversity loss. Through the Convention on Biological Diversity, countries have recognized the dual importance of protecting their natural heritage and using their natural resources in a sustainable

E. D. Wiafe, *Wildlife Conservation and Agriculture Development in Africa*, Sustainable Development Goals Series, https://doi.org/10.1007/978-3-032-09233-5_6

manner. The impact of past unsustainable use of resources has led to a large growth in conservation management. With input from conservationists, biological scientists, ecologists, and many other professionals, the world community has now gained a greater understanding of what is required to protect species, maintain ecosystem health, and preserve genetic diversity.

Wildlife conservation is the practice of protecting plant and animal species and their habitats. Our world today is inhabited by nearly 2 million known species and it is in need of being conserved. Many of our nations' resources that are used to produce food, energy, and industrial products are provided by plants and animals. Wildlife can be severely damaged by the activities of people. Illness is introduced by domestic animals, people collecting plants, and general land development have caused the decline of wildlife in many regions. Many are concerned about the high rate of extinction for many plants and animals.

6.3 Challenges Faced in Wildlife Conservation

Major declines in the range and abundance of species lead to genetic loss, reduced ecosystem function, and a loss in the ecosystem services that wildlife provides to people. Live animals are also an important resource, with many being captured and sold on the local and international market. Although there is a domestic trade in Africa for pets, the primary destination for wild-caught animals is the pet and zoo industry in the USA, Middle East, Europe, Japan, and increasingly, China. Capture methods can be highly damaging to populations and ecosystems, and many animals die in the process or in transit to their destination. This results in depleted populations of certain species and reduced biodiversity. The international market for wildlife has also led to an increase in alien species introductions as exotic pets and "beefed up" game animals are often released into the wild, out-competing native wildlife.

High-value species are often specifically targeted, for example, elephants for their ivory and rhinoceros for their horns. This can lead to the local extinction of a species, an example being the northern white rhino, of which there are now only three individuals left on earth. This is merely the extinction vortex at which the last remaining individuals are most vulnerable. At this point, there is little probability of long-term survival of the species, and it is often referred to as a "living dead" situation.

Commercial hunting for wildlife products, including skins, hides, tusks, and body parts, is essentially a resource-extraction industry. Wildlife meat is also an important resource in marginalized areas. This can be explained by its high protein value in relation to livestock. In East and West Africa, for example, it is often considered the preferred meat.

Wildlife crop-raiding is another issue in agricultural areas and can lead to human–wildlife conflict, often resulting in the killing of "problem" animals. Unsustainable off-take from these activities inevitably leads to declines in a species' abundance in a given area, resulting in local or complete extirpation of that species.

From the past to the present, numerous threats to Africa's wildlife, which can be considered a series of challenges and tests, have weakened the continent's biodiversity. The two main factors for these threats historically have been the demand for wildlife products and human population growth. These factors are still, to this day, the greatest threats to Africa's wildlife and home to various species.

Considerable challenges are faced in wildlife conservation in Africa. The continent's wildlife resource is immense, rich and varied. It includes both land and marine resources and is widely recognized as holding an international heritage. The extent of these resources is huge, with some 200,000 different species, many of them not found anywhere else in the world.

6.4 Wildlife Conservation Strategies

Protection of habitat often involves the development of ecotourism. This is generally viewed as an environmentally friendly industry and a sustainable way of making use of natural resources. Showcasing the natural beauty and ecosystems of a given area can encourage its preservation in the face of encroachment by human development, and the tourism revenue can provide a strong incentive to preserve the natural state of an area. Because of the emphasis on preservation, the potential long-term effects of ecotourism on wildlife are an area of concern. However, when properly managed, it has been held up as a conservation success story, as is often the case in Southern Africa.

At the broadest level, many wildlife conservation strategies can be classified into two principal categories: those that are aimed at protecting the wildlife and those that are aimed at protecting the habitat. Techniques that are aimed at protecting the wildlife include establishing protected areas, such as national parks and other wildlife reserves. But legislation alone is rarely successful in the face of active land development, and poaching has devastated many areas set aside for wildlife. In many cases it is necessary to take a direct hand in protecting the wildlife, by conducting anti-poaching patrols such as aerial surveillance, and sometimes controversially by establishing a designated area in which a severely depleted wildlife population can be sustained, at the expense of the local human population. Wildlife on private land can be preserved through the establishment of private wildlife reserves, onto which no hunting or development is allowed. Often these reserves will be established by landowners who are primarily interested in conservation, or who see more long-term benefit in the sustainable use of their land.

6.4.1 Protected Areas and National Parks

Protected areas are the basis for all conservation activities and act as sanctuaries for the preservation and maintenance of biodiversity and associated cultural resources. They range from strict nature reserves to multiple use areas, with the majority of category finds of most benefit to conservation—recognizing natural resources are both finite and irreplaceable. No matter what the category, all protected areas have the same basic principle: to maintain the sustainable health of the environment and to provide a foundation for recovery of biodiversity and local livelihoods, integrating conservation and human development. Benefits can be direct or indirect and can be social, economic, environmental, cultural, or educational, improving the lives and environments of all those involved. For example, in Uganda's Bwindi Impenetrable National Park, the Mountain Gorilla Project receives 20% of the funds generated from tourism in the local area, which is enticing the majority of the funding to build more schools and hospitals—improving the general health and education in the area. Gorilla tourism is one of the many revenue-generating activities taking place in protected areas. It is an industry growing in importance within conservation all over the world. In Africa, tourism in parks and reserves has been an off and on activity since colonial times and the involvement of the local community has changed greatly over the years. Now there is an emphasis on how revenue generated can be ploughed back into the local economy and provide the people living on the edge or within the parks, tangible benefits. This has led to tourism agreements and development of resource use in which the local population is involved. An example is in Benin where hunting was once very damaging in Pendjari National Park, it is now becoming one of the most successful ventures in community-based conservation in the region, and game populations have increased due to effective law enforcement and anti-poaching.

6.4.2 Community-Based Conservation (CBC) Programs

Today, CBC projects have grown in both scope and number. They are defined as partnerships between wildlife and communities that result in reduced threats to the survival of wildlife and resources for the community. The most salient feature of these programs is that, through a series of agreements, communities are persuaded to set aside land for wildlife conservation in exchange for tangible benefits. This approach displays an opportunity cost to conservation. The Village Game Scouts program operates in this way, in which specific benefits are given to the village as a result of a decrease in poaching levels.

The Maasai Steppe experience and the prelude to game manage ranch facilitated concrete changes in resource control. Although in the long term interest of future generations of villagers, the planning process facilitated the single most important change of recent times in a Kenyan village's decision making. It enabled the villagers to obtain grazing leases from Uaso N'gira Safari Ranch and left the road and water projects in the hands of the village leadership, with the government assuming only the role of provider. This same process set a precedent for decisions regarding land use and meant an increasing acquisition of land use information. Aware of the potential for wise land use decisions based on accurate information, Chip Wilson pondered whether a wilderness area, once it had been inventoried and mapped, could serve as a strategic resource of information and training for communities involved in land use planning.

In the brief period in which Maasai Steppe operated, local leaders demonstrated a keen interest in acquiring accurate information on the status of land use and the potential sites for valley dams and terracing. During the micro catchment land use planning exercise, village leaders consulted informally among themselves, and later as a group, to establish priorities and make plans. The planning process, culminating in the meeting with their district councillor, constituted fundamental political organizing by the village leadership initiative not seen for some years.

Community-based conservation (CBC) programs have been initiated as a result of the failure of fortress conservation to achieve its goals. These programs emerged in the 1980s, at a time when global awareness of conservation and the need to involve local communities was high. The first project, called the Maasai Steppe, was an integrated rural development and nature conservation project. It aimed to solve land use problems without damaging the environment. Its key principle was for land to be set aside for conservation purposes, and community development and land use should be based on accurate and up-to-date information.

6.4.3 Anti-Poaching Efforts

Illegal hunting has wrecked a lot of damage on wildlife throughout the world, and to many species and different ecosystems. Poaching can cause extinction in a species, and have an impact on a species' population. The rhino and tiger are two species that have been hit by poaching fairly bad. If the problem is to be stopped, effort is needed on law enforcement, community involvement, and student education. There are many organizations meant to help conserve wildlife and put a stop on poaching through law enforcement. For example, the Forest Rangers Society is an Indian-based society. Based in India, it focuses on forests society and the various issues that are related to the Indian forests. This would be dealing with every issue right from tiger extinction and camel conservation. This society covers all wildlife, plants, and trees to have a better future for the generations to come. Another anti-poaching organization includes the Wildlife Police in South Africa run by the Endangered Wildlife Trust. This is a small, elite, highly trained force that is adequately equipped to fight wildlife crime throughout South Africa. This does sound like a good idea and a step in the right direction towards banning poaching and affecting the economy of the respective countries.

6.4.4 Wildlife Monitoring and Research

Research is also critical in shaping effective conservation policy and practice. Often conservation and management decisions are based on inadequate information, which may lead to wasted resources and ineffective policy. By investing in research, which seeks to answer specific management questions, there is a far greater likelihood of achieving the desired conservation outcome. Much of the research required to inform conservation management is not being pursued by universities or government agencies. It is therefore important that on the ground conservation organizations are equipped with both the funding and skills required to undertake research that is directly relevant to their conservation goals.

In order to achieve successful wildlife conservation goals, it is necessary to monitor the status and trend of wildlife populations. Wildlife monitoring is essential in assessing the effectiveness of management and conservation policies. The evaluation of available data provides the best scientific foundation for future conservation efforts.

6.5 Success Stories in Wildlife Conservation

Species recovery programs have been a resounding success. One of the dilapidated and frequently failed Namibian conservation programs has managed to maintain the world's largest population of the critically endangered black rhinoceros. Due to an increasing demand in rhino from other African nations, where rhino populations have been severely depleted, a rhino custodian program has had huge success in translocating rhino to communal land in order to increase the range and numbers of the species. Similarly, a South African cheetah recovery program has had success in both captive and wild cheetah recovery efforts, boasting increases in numbers and subsequent relocations to small fenced reserves, the first step to wild cheetah conservation.

There are a number of successes in wildlife conservation in Africa—in spite of the huge tasks and tall orders being set and met, in what sometimes seems an insurmountable challenge to the many threatened and endangered species. These successes can be seen as shining examples of hope in a region of the world reeling from poverty and other economic and social conditions that seem to constantly take precedence over the environment as a priority in national and international agendas.

6.5.1 Species Recovery Programs

Somewhat paralleling the increasing threat of loss of species, many proactive recovery programs for species and population recovery have been initiated. Recovery is the process by which decay of an endangered species is arrested or reversed and the process which eventually leads to delisting of a species. Essentially, a plan to make a recovery plan aims to ensure the long-term viability of a target population. The United States Endangered Species Act has laid the foundation for recovery planning for any listed species. Using the conservation biology approach to recovery planning ensures the survival of a species. In a genetic perspective, captive breeding and restoration ecology are often involved as well as direct manipulation of a population (proven successful for instance with elephant seals) and manipulations at the ecological levels are aimed towards enhancing the resilience of the species to critical threat factors. The Costa Rican Gold Frogs is one example of a species that has a detailed recovery plan. An alliance between three organizations in Costa Rica are conducting its recovery which includes propagation in captivity, research on the pathogen (Chytrid fungus) that led to its decline and reintroduction in areas free from these pathogens.

In Africa, an excellent example is the African Rhino Specialist Group which has produced a detailed "Rhino Conservation Programme" which included recovery planning for both black and white rhinos using varying methodology across different range states.

6.5.2 Conservation Education and Awareness

Better-known conservation approaches in Africa include the very interesting "Camps International," which is a non-governmental organisation that focuses on community-based conservation. However, this is achieved through sending expeditions of school students to third-world countries to help benefit the people and wildlife. One of their examples is the initiatives with local Maasai communities to help secure a future for Maasai people and wildlife. They support the idea of education as a source of social change and wish to provide young Maasai with a western education whilst preserving their culture and knowledge. This is to ensure that the young Maasai will be able to rationalize the differences between their traditional beliefs and practices and the changing world. The education provided gives Maasai children the chance to secure new livelihoods in an increasingly modern and competitive world—as a result, they are less likely to be forced to exploit wildlife and wild lands to survive. Ignorance is therefore combated and it is hoped that less and less people will resort to subsistence farming and cutting of forests to maintain their livelihoods. This is echoed in other parts of Africa, which is imperative on educating local peoples and children to gain their support for conservation and an understanding of the need to preserve wildlife. In higher-profile decisions on wildlife, education remains a powerful tool for political change. For example, the Jane Goodall Institute (JGI) Roots and Shoots is an environmental and humanitarian program for young people that can be found in over 130 countries. Its mission is to foster respect and compassion for all living things, to promote understanding of all cultures and beliefs, and to inspire individuals to take action to make the world a better place for people, other animals, and the environment. This has already been shown to influence several international conservation decisions. We have seen that a lack of education is a root cause of conservation issues. It is now necessary to look at the long-term investments towards conservation education and how these initiatives support those in Africa actually working in conservation. This comes under an agreement that there is a dominant need for African biologists and for conservation biology to be added as a professional discipline throughout Africa. Future progress of conservation in Africa will depend on the human capacity to care for and manage the continent's biological resources.

6.5.3 Sustainable Tourism and Conservation

One of the best-known examples of creating a link between tourism and conservation is mountain gorilla tourism in Rwanda, and Uganda. Mountain gorillas are directly protected from poachers within the Virunga national park, as increased tourism provides revenue for the park warden and staff to better patrol the gorilla sector and enforce anti-poaching laws. Live Gorilla groups are also tracked by government organised groups of tourists who pay a fee to the park, and all visitor fees create substantial revenue, one third of which directly goes to the park. This money has in turn funded many infrastructure and community projects and a significant portion has been invested into gorilla and other wildlife conservation outside the tourism sector. An examination of 12 years of gorilla census data in the Virunga has shown a 26.3% increase in population and distribution of gorillas in the sector, and no doubt increased tourism has played a large role in this trend.

The final area I would like to cover, and one of the most valuable ways of maintaining world heritage through supporting conservation, with the added justification of satisfying public interests is through supporting wildlife and habitat, be it in a rainforest or savannah. The work of several primate focused projects intertwines with ecotourism at a number of levels. In Uganda for example, the Kibale Fuel Wood Project focused on primate habitat conservation around Kibale National Park, works closely with the Uganda Wildlife Authority and Uganda Safari Lodges to classify and develop a tourism circuit around the park, which visits key primate viewing areas. At a similar level, the World Wildlife Foundation in the Central African Republic works with Dzanga-Sangha protected areas to promote tracking of lowland gorillas and elephants and has developed a successful tourist research station. While the overall effect of tourism at a local level can often be damaging through habitat destruction and increased poaching, the development of tourism circuits in protected areas can be very beneficial for wildlife protection in the long run.

6.6 Future of Wildlife Conservation in Africa

A key requisite for successful integration of conservation with broader rural development is a comprehensive land-use planning process, in which local people are involved in defining options for the future of their areas, and where these choices are weighed in terms of the likely impact on people and the environment. This is essential for the maturation of community-based conservation from a mere slogan to a meaningful reality. By far the most common form of land use planning in Africa is the Regional Resource Inventory (RRI) exercise. These exercises have achieved mixed success in terms of direct conservation outcomes, but they remain an excellent opportunity for integrated conservation and development initiatives to be promoted at national and regional levels. It is crucial that wildlife and biodiversity concerns are central to RRIs, as they provide the only productive argument against further degradation and loss of habitat to agriculture and livestock. Adequate funding and technical support will be required to ensure this.

The future of wildlife conservation in Africa is at a critical juncture. There is growing recognition among the countries of sub-Saharan Africa that conservation has to be integrated with development if wildlife is to survive. In the past, most conservation in Africa has been carried out by foreign NGOs or donor organizations, and a primary motivation for conservation by national governments has often been the attraction of tourism revenue or the receipt of development aid. This approach is no longer feasible. Foreign funding for conservation will likely diminish because of the ongoing global economic downturn and the escalating cost of anti-poaching measures in the face of expanding human populations. Therefore, it is increasingly important to closely link conservation with the poverty alleviation and public health agendas in Africa, and to build conservation capacity and expertise within the continent.

6.6.1 Emerging Technologies in Conservation

Its potential was realized in the late twentieth century and sparked numerous projects to investigate the extent to which remote sensing can be used to monitor wildlife and the environment. NASA's moderate resolution imaging spectroradiometer (MODIS) has been the focus of remote sensing development in terms of conservation and development of products, which are designed for practical uses in management and decision making. Although it is not a conservation specific tool, it has been used by a multitude of conservation biologists and has potential to address a number of issues relevant to African conservation including detection of poaching and changes in land cover for agriculture.

Remote sensing is a powerful tool that has already been adopted as part of conservation management in Africa. It involves the collection of information about the environment via sensor devices that are not in physical contact with the object under investigation. A myriad of technologies including aerial photography, and various types of radar and satellite systems can be used to monitor land cover/land use change, deforestation, fire, water quality and movement and changes in wildlife and their habitats.

A paradigm shift is discernible with the advent of emerging technologies, which are proving to be invaluable in terms of improved monitoring and surveillance, more accurate data collection, as well as analysis and assessment of conservation programs. The World Wide Fund for Nature (WWF) has recognized the potential benefits of technology in achieving conservation goals and is actively exploring how technological applications such as remote sensing, geographical information systems (GIS), and satellite imagery can be used to gain more insight about key processes and threats in regions of interest.

6.6.2 Collaboration and International Partnerships

One of the methods that has proven to be effective and is believed to continue to be so is through collaboration and international partnerships. Particularly in Africa, where in many cases the host country does not have the financial or technical resources to effectively manage their wildlife resources. Historically, many African countries have formed international bilateral agreements with European countries to provide funding and support for the management of wildlife. However, these arrangements have often been to the benefit of only a few species or national parks. The future of such agreements it is hoped will be more focused on enabling African countries to sustainably manage and utilise their wildlife resources in a manner that is consistent with their development aspirations. Another still developing form of partnership has been the creation of debt-for-nature swaps. These agreements, usually between a conservation organizations and a national government, or sometimes a commercial enterprise, involve the purchase at a discounted rate of large portions of a country's foreign debt. This debt is then legally converted into a long-term commitment to fund conservation activities in the debtor country. Although still in the early stages and sometimes criticized, it has the potential to release significant funding for conservation at no cost to the donor country. An example was the 2001 agreement between the government of Seychelles and a commercial enterprise, which for the purchase of approximately US$ 3 million of Seychelles debt, committed around US$ 8.75 million to a trust fund for the conservation and management of the Seychelles environment.

6.6.3 Climate Change and Its Impact on Wildlife

Climate change is likely to affect wildlife and its habitat in profound ways. For example, the Wildlife Conservation Society (WCS), predicted the impacts of climate change on the habitats of over 3500 species in African biodiversity hotspots. They found that in the Albertine Rift area, an increase in average global temperatures of 3°C would have very significant effects. The area of suitable habitat for the endangered African elephant would decrease by 39%, and the ranges of 24 protected area network sites would decrease by an average of 42%. The continuum of effects across the region included increased habitat fragmentation, disrupted ecological and evolutionary processes, and a loss of site-level conservation investments. The predicted effects on forest ecosystems in the Albertine Rift were linked to clearing of areas for farmland often by refugees from civil conflict and other human pressures. This underlines the importance of integrating climate change impact assessments into conservation planning and management so that steps can be taken to reduce the future overall impact of habitat loss on wildlife. An analysis across sub-Saharan Africa as a whole suggested that 70% of protected area sites have an increased probability of major changes to the species present, and the tree cover or biome type can be expected to change at 60% of sites. The most severe effects on protected areas and their inhabitants are likely to occur in East and West Africa.

6.7 Conservation Partnerships in Africa

6.7.1 Collaborative Efforts

An increasing number of conservation partnerships are forming between the private sector and various conservation organizations. This trend stems from the realization that sound environmental and natural resource management is crucial to several industries such as eco-tourism and outdoor apparel. Partnering with conservation organizations allows businesses to operate in a sustainable environment and ensure the existence of resources that they depend on. A good example is the recent partnership between Conservation International and Shell corporation. Although there has been much scrutiny over this relationship, it is argued that engaging multi-national corporations in conservation efforts will yield greater benefits than attempting to fight or ignore the impacts they have on the environment. This is especially true in Africa where the rapid expansion of industry and natural resource extraction can have detrimental effects on the environment if not done responsibly.

Local communities and conservation groups have also found themselves entering into conservation partnerships. Perhaps one of the most renowned forms of collaboration is the Communal Areas Management Program for Indigenous Resources known as CAMPFIRE. This program was designed to devolve rights to the use and management of wildlife resources to communities located in and around protected areas. With assistance from several non-governmental organizations, CAMPFIRE has found backing from several donors and has proven to be an effective way to link rural development with wildlife conservation in Zimbabwe and other countries in Southern Africa.

The collaborative efforts of conservation partnerships take place on many levels, often including government and non-profit organizations such as the United States Fish and Wildlife Services and the African Wildlife Foundation. The programs that are possible through such partnership efforts help to raise the standard of resource management in many of Africa's protected areas. Conservation partnerships help to address the critical needs for effective management and have led to both policy initiatives and lasting on-the-ground accomplishment (Ponte et al. 2021; Lindsey et al. 2021; Cadman et al. 2020; Zhao et al. 2024).

6.7.2 Government and Non-Profit Organizations

WWF can also work with government indirectly through third-party organizations, such as IUCN's West and Central Africa Programme. In Cameroon, CIDA has helped government organizations, to date the Ministry of Environment and Forests and the Ministry of Agriculture Rural Development, implement their Enable and Iterate Agriculture Policy Framework for land use change and forestry alternatives. The program aims to conserve biodiversity by reducing the conversion of extensive cropland.

In Belize, WWF has been working with the Ministry of Forestry to develop a joint program with the United Nations, aimed at reducing deforestation in Central America. This program seeks to address deforestation and forest degradation in the region by promoting more sustainable land use, profitable rural development, and alternative land tenure arrangements.

WWF has focused recent efforts on gaining a better understanding of the role of land and resource tenure, as well as land use change patterns within countries. We emphasize the conservation of remnant natural habitats that are currently well protected but at risk due to conflicting plans. This work has led to several key partnerships with government agencies and NGOs, including The Nature Conservancy and the Stable Future for PNG Programme.

Conservation organizations may work with government organizations in many different capacities, ranging from partnership on a single project to implementation of a larger program or policy shift. While much of WWF's work with government occurs at the local level, we also engage in policy work with national governments, often forming multi-year partnering agreements or memoranda of understanding. The goal is to improve the effectiveness of government in implementing conservation-friendly policies, to identify and fulfill critical conservation gaps, particularly in areas located on the edges of protected areas and reserves (Nohlová 2020; Filer et al. 2020; Middleton et al. 2020).

6.7.3 Local Communities and Conservation Groups

By enabling the people to derive direct benefit from wildlife and land, communities are encouraged to set aside land for the purpose of wildlife and natural resource management. This has been achieved largely through the devolution of user rights to communities over areas now designated as wildlife management areas or community forestry projects. In Zimbabwe and Zambia, this has been synonymous with the CAMPFIRE project and in Namibia the establishment of conservancies. In some instances, communities have formed their own conservation and/or natural resource management organizations to help facilitate the process. Here are two summaries of how community conservation works.

The partnerships between local communities and conservation groups are crucial to both the people and the environment of Africa. It is widely recognized that the indigenous people of Africa, who are the true owners of the land, have been systematically excluded from the benefit of tourism and land use. This has led to a situation where the only economic benefit they see from wildlife and wild lands is the damage done by crop raiding elephants, disease-spreading wildlife and human–wildlife conflicts (Dawson et al. 2021; Musavengane and Kloppers 2020).

6.8 Private Sector Engagement

While the potential benefits of such partnerships are sizeable, there are also significant risks to both the reputation of the conservation organization and the private company involved. In general, it is recognized that it is important to engage with all companies whose operations have a significant impact on the environment, in order to steer them towards sustainability. However, there is a fine balance between engagement and the risk of being co-opted by the company, or being seen to endorse an unsustainable activity. This has been a key issue in discussions on the best way to influence the extractive industries. An IUCN review of experience with engagement of the private sector in sustainable development emphasized the need for a long-term strategic approach, rather than forming ad hoc partnerships and reacting to approaches from the business sector. The review also highlighted the need for a participatory process that involves not only the company management, but all stakeholders affected by the company's operations. Such a process can often be at odds with the expectations of either party in a partnership with a conservation organization.

In recent years, the private sector has recognized the importance of biodiversity and ecosystem services in Africa for their own sustainability. For example, Coca Cola has invested in water conservation around their sugarcane farm in Gaole, while Dow Chemical has funded a development project in the Bana district of Cameroon. The trend is for such investment to be channeled through third parties, be they NGOs (for example, the Dow project in Cameroon is managed by the World Wildlife Fund) or through bilateral relationships. At the same time, conservation organizations are seeking out partnerships with private companies to tap into their resources, funds and expertise. For example, UNEP has recently signed an agreement with Terra Choice to implement a global program to improve the environmental and social standards of tourism. A number of these initiatives are documented in case studies in the IUCN/IFC publication "Doing Better Business for Nature" (Dawson et al. 2021; Abukari and Mwalyosi 2020; Armitage et al. 2020).

6.9 Public-Private Partnerships

The third tier is aimed at business sectors and it involves the development of Best Practice Guidelines for industry sectors on environmentally friendly and sustainable development of their business. This program provides a framework for business community involvement in conservation and encourages individual and collective conservation initiatives.

The second tier provides a network for like-minded businesses to come together to share experiences and strategies in working towards common conservation and business objectives. This network will provide support for businesses to develop their own conservation strategies and will also facilitate joint projects between partnering businesses.

The first tier of the partnership provides a mechanism for individual businesses to assist conservation by supporting specific conservation projects. In return, WCPA will provide advice and assistance to these businesses in implementing environmentally friendly policies within their business operations.

This partnership model encourages business communities to develop and maintain a sustainable environment through a three-tier partnership program. The goal is to encourage businesses to adopt a culture of corporate environmental management and to integrate conservation of biodiversity into their core business.

6.10 Benefits and Challenges

Co-management arrangements, like conservation partnerships, have proven to be a successful conservation tool in southern Africa and other developing regions. The WILD Act's provision recognizing the benefits of public-private initiatives is a good starting point for more fully realizing the potential of conservancies and other co-management arrangements. There are myriad potential benefits to both the public and private sectors and civil society (local and international) through such partnerships. For government agencies of developing nations, where funding for conservation is chronically inadequate, cost-sharing arrangements with the private sector can provide a means to escape the "fiscal strait-jacket". Depending on the amount of government funding already allocated to protected area management, matching grants from private organizations could free up additional public sector funds for conservation activities, potentially even leading to a situation demonstrated in Namibia, where government cost-recovery in a conservancy area actually generates revenue for other protected areas. Though cost sharing is typically portrayed as a 50–50 split between public and private stakeholders, a variety of creative funding mechanisms have been employed to support everything from infrastructure development to community-based natural resource management. From tourism levies in Zimbabwe to conservation hunting programs in Tanzania, conservancies and joint venture agreements have provided a testing ground for new methods to link protected area viability and financial sustainability. Whether implemented at project, site, or national level, initiatives to build human resource capacity for conservation are envisioned to be an integral component of the overall sustainable development agenda for various states. A percentage of newly generated funds, or earmarked public sector funds, could support various types of training programs to fill the gap of social and natural science capacity eliminated during the 1980s structural adjustment period seen in many countries.

6.10.1 Examples of Successful Partnerships

1. **Great Limpopo Transfrontier Park (GLTP)**
   Partners: South Africa, Mozambique, Zimbabwe, and several international conservation organizations.
   Overview: GLTP is a transboundary conservation area covering over 35,000 square kilometers, connecting Kruger National Park in South Africa, Limpopo National Park in Mozambique, and Gonarezhou National Park in Zimbabwe.
   Achievements:
   Promoted free movement of wildlife across borders, enhancing genetic diversity.
   Strengthened anti-poaching efforts through coordinated ranger patrols.
   Increased ecotourism opportunities, benefiting local economies.
2. **The Lewa-Borana Conservancy, Kenya**
   Partners: Local communities, private landowners, the Kenyan government, and international donors.
   Overview: A private-public partnership protecting critical habitats for rhinos, elephants, and Grevy's zebras.
   Achievements:
   Successfully increased black rhino populations through intensive protection efforts.
   Supported community development projects, including schools and healthcare.
   Created sustainable tourism models that provide income for local residents.
3. **Gabon's National Parks System**
   Partners: Gabonese government, international NGOs like the Wildlife Conservation Society (WCS), and UNESCO.
   Overview: Gabon established a network of 13 national parks covering 11% of the country's land area to protect forest elephants, gorillas, and other wildlife.
   Achievements:
   Reduced illegal logging and poaching in critical areas.
   Developed ecotourism initiatives that contribute to Gabon's economy.
   Established robust research and monitoring programs for biodiversity conservation.
4. **The Southern African Wildlife College (SAWC), South Africa**
   Partners: Peace Parks Foundation, WWF, and South African National Parks (SANParks).
   Overview: A regional training institution offering conservation education to practitioners across Africa.
   Achievements:
   Trained thousands of rangers and conservation professionals in wildlife management.
   Implemented community-based conservation initiatives to reduce human–wildlife conflict.
   Strengthened capacity for anti-poaching and habitat management across southern Africa.
5. **Okavango Delta Conservation Project, Botswana**
   Partners: Botswana government, local communities, and NGOs such as Conservation International.
   Overview: A collaborative effort to protect the Okavango Delta, a UNESCO World Heritage Site and critical habitat for elephants and hippos.
   Achievements:
   Implemented sustainable water management practices benefiting both wildlife and communities.
   Increased community participation in conservation through tourism ventures.

Reduced poaching incidents through strengthened enforcement and monitoring.

6. **Mountain Gorilla Conservation, Virunga Region**
   Partners: Rwanda, Uganda, DR Congo, Dian Fossey Gorilla Fund, and the International Gorilla Conservation Program (IGCP).
   Overview: Conservation initiatives in the Virunga Mountains, home to critically endangered mountain gorillas.
   Achievements:
   Increased mountain gorilla populations, with numbers surpassing 1,000 in recent years.
   Boosted ecotourism revenues, contributing to local economies and funding conservation efforts.
   Fostered cross-border collaboration in law enforcement and research.
7. **The Niokolo-Koba National Park Conservation Project, Senegal**
   Partners: Senegalese government, African Parks, and UNESCO.
   Overview: Focused on reviving the park's biodiversity, which includes lions, chimpanzees, and African wild dogs.
   Achievements:
   Reintroduced critically endangered species like the West African lion.
   Engaged local communities in park management and decision-making.
   Improved park infrastructure and anti-poaching strategies.
   These partnerships illustrate the potential for multi-stakeholder collaboration in addressing conservation challenges, benefiting both wildlife and communities across Africa.

6.11 Role of Government in Facilitating Partnerships

The policy development process, which began in 2004, has involved the engagement of stakeholders at all levels and has culminated in in-depth discussions on each of its 22 draft policy statements. This process has demonstrated participatory democracy in action, and when the policy is ratified, it will ensure government support for the CBNRM strategy in the future.

The policy involves a partnership between NACSO and the government that aims to: (1) Create an enabling environment for rural communities to manage their natural resources on a sustainable basis by taking wildlife and tourism out of the commons, thereby reducing the tragedy of the commons. (2) Strengthen rural communities' rights with respect to the ownership and management of wildlife and other natural resources. (3) Provide greater national and international awareness of the rights and achievements of rural communities in managing and conserving their natural resources.

In Namibia, the government has given communal conservancies rights over wildlife management and tourism on communal land by passing the Communal Conservancies Act in 1996, its enabling legislation. This has led to the establishment of over 29 conservancies and the formation of the Namibian Association of CBNRM support organizations (NACSO). This umbrella organization has received government support in the past, and its role is to provide a forum through which the government and other stakeholders can develop and review conservation and natural resource management policies affecting communal land. This has resulted in NACSO being given the mandate to coordinate the development of a national policy for Communal Area Resource Management (FORM) (Wenborn et al. 2022; Kalvelage et al. 2021; Siyambango and Enkono 2022).

Conservation partnerships supported by government agencies, such as those existing in Namibia and Zimbabwe, can be particularly effective. These partnerships are typically between the government and rural communities and involve the devolution of user and management rights over wildlife and natural resources to local communities. These rights are granted through mechanisms that allow local communities to benefit from managing these resources sustainably.

6.12 International Organizations

The World Bank and the African Development Bank have been working with several African nations to assist in the economics of environmental management. This work is focused on policies, research and development, and integrative planning aimed at sustainable resource use and poverty alleviation. The World Bank has also been involved in a number of "Global Environmental Facility" (GEF) projects and programs in Africa. These programs are usually focused on specific issues such as biodiversity conservation, international waters, climate change, land degradation, the ozone layer, and persistent organic pollutants. The primary objective of GEF is to support developing countries in meeting the terms of international environmental agreements and to do so through the integration of these objectives into national development plans. The GEF is implemented through UNEP, the United Nations Development Program (UNDP), the World Bank, and various development agencies.

The United Nations Environment Program (UNEP) is active in working with African nations in the field of environmental conservation. UNEP has a Regional Office for Africa, and its work in Africa is coordinated by the Division of Regional Cooperation. UNEP's work in Africa is focused on six key areas: supporting environmental sub-programmes and assessments, early warning and emergency response, mainstreaming environment for sustainable development, policy and law, resource efficiency, and environmental governance.

International Organizations are key actors in assisting African governments to build and maintain a healthy and sustainable natural environment. These organizations help African countries in environmental planning and to create and implement policies and regulations designed to conserve natural resources and the environment. A number of international organizations are also actively involved in the conservation and development process throughout the African continent.

6.13 United Nations and Conservation Initiatives

The United Nations Environment Programme (UNEP) was established in 1972 as a result of discussions on the need for a global agency to coordinate environmental preservation. The idea of UNEP was conceived at the Stockholm Conference, resulting in the establishment of an intergovernmental body that would provide the focus for global environmental activities. Since its inception, UNEP has placed increasing focus on global biodiversity conservation, recognizing it with the establishment of a dedicated Biodiversity Programme in the Division of Global Environment Facility Coordination in 1990. A key event was the setting up of the World Conservation Monitoring Centre, as a joint initiative with 3 collaborating organizations. In the past 5 years, UNEP has set in place a series of Inter-Governmental Forums in which different countries seek consensus on pertinent international environmental issues and a work programme aimed at protecting the global environment. In particular during the 3rd Meeting of the Intergovernmental Panel on Forests, a decision was made to develop a legally binding agreement on forests and to establish a new international body to monitor and oversee the management sustainable development of the world's forests. During its history, UNEP has had successful projects in the form of funding protected areas, implementing international agreements and work in fields such as environmental law. These projects have been achieved through a series of mechanisms such as funding from the World Conservation Monitoring Centre in Cambridge, UK, a collection of 32 institutions which works to provide information for enabling the monitoring and conservation of the Earth's biodiversity and the UNEP-GEF (Global Environment Facility). UNEP's work with the GEF has been especially significant. An agreement was signed in 1991 enabling the GEF to operate as the funding mechanism for UNEP's international waters and the Montreal Protocol on substances that deplete the ozone layer. This was then extended to cover the areas of biological diversity and climate change. The funds provided for the Montreal Protocol were crucial in assisting developing countries to comply with the agreements and phase out substances that were having adverse effects on the environment. UNEP has also worked on projects alongside the secretariats to some of the conventions it has endorsed. An example of this would be the project with the Convention of Biological Diversity which resulted in the Caribbean Biological Corridors Project. The objective of this project was to design and support implementation of a set of activities to alleviate threats to biological diversity in the Caribbean through conservation and sustainable use of its biological resources (Desai 2020; Canton 2021; Ivanova 2020).

6.14 Role of World Bank and African Development Bank

In the period since the 1992 Rio Summit, the Bank has re-evaluated the role of its non-lending services such as economic and sector work and policy advice. This has been encapsulated by the concept of the "knowledge bank" and the Bank has invested considerable resources in producing and disseminating research findings. During the same period, there has also been an increased engagement with civil society and the private sector. The relevance of these changes for conservation in Africa is difficult to assess at present, although the Bank's increased focus on environmental sustainability and poverty alleviation should create a favorable environment. This would suggest that conservationists might seek in the future to make more use of the Bank's economic and sector work and policy advice.

The role of the World Bank and the African Development Bank is major in terms of funding conservation efforts within the African continent. A variety of regional and sub-regional projects have been supported (usually with GEF assistance) in an equally diverse range of countries. Projects range from those directly dealing with the protection of biodiversity, through institutional and capacity-building initiatives, to attempts to ameliorate conservation-threatening problems outside the conservation sector. All of these have an impact on the prospects for conservation in Africa, while the Bank's corporate emphasis on poverty alleviation and sustainable development means that its involvement in natural resource management is likely to increase in the future (Michaelowa et al. 2021; Ofori et al. 2022; Koloma 2021).

6.15 Collaboration with Non-Governmental Organizations

In recent times, the World Bank and the African Development Bank have shown much interest in what non-governmental organizations are doing towards conservation in Africa. This has led to an increase in loans and project funding from those two establishments towards government and community-based conservation projects. This indirect funding has helped to increase the capacity and effectiveness of non-governmental organizations at the local level. An example of such co-financing of conservation projects in Africa can be seen with the African Wildlife Foundation. This organization works to protect wildlife and wild lands in Africa through the use of innovative strategies and an inclusive approach which involves local people. With funding from the World Bank, the African Wildlife Foundation has been able to identify critical land use strategies and to design rural development projects that address the needs of both the environment and poverty-stricken rural Africans. This has contributed to the building of sustainable livelihoods for rural people and the reinforcement of conservation, which has often been a sidelined secondary priority on communal lands. The World Bank in this case has invested in the future of rural Africans and their environment, as the lessons learned and the tools developed are applicable to models that can be replicated across Africa and used in various conservation and development scenarios.

6.15.1 The Role of Research

A major incentive for further research has to be the capacity to change policies and interventions so that they are more beneficial to both people and the environment. In a recent article, Kremen called for greater synthesis and availability of research findings to inform land-use planning and prevent the many detrimental environmental impacts of ill-informed development in the tropics.

The importance of biological and socio-economic research to conservation practice cannot be overstated. In a recent international survey, over 300 conservation biologists identified priority topics for research and then evaluated the relative importance of these topics to the conservation of biodiversity. The results suggest that some of the most important questions for conservation remain largely unaddressed. For example, a review of the ecology of the (*Gorilla gorilla*) as a basis for conservation action identified only one relevant study throughout Central Africa over a 13-year period. This contrasts with other areas of research where the scientific understanding far exceeds the information necessary for effective action. For example, the same review listed 43 empirical studies of snares and their effects on wildlife in Central Africa, despite the widespread use of this hunting tool and a clear ability to infer general patterns from far fewer studies.

Conservation is a knowledge-based activity, and research is a fundamental tool for understanding the processes essential to the maintenance of biodiversity and the sustainable utilization of resources. Research can take a variety of forms, including formal scientific investigation, surveys of local knowledge, or ecological monitoring using indicators of environmental change. Research projects can be initiated by scientists, resource managers, or communities and may be directed at increasing fundamental knowledge, solving specific problems, or informing the development of new policies.

6.15.2 Scientific Studies and Data Collection

A wide variety of scientific investigations and research studies are relevant to conservation, ranging from genetic and reproductive studies that elucidate the biology of a species, to economic analyses of human incentives for resource use. Underpinning them all is a need for accurate and relevant data. Too often projects have failed because an inadequate understanding of the biological, social or economic factors critical to the problem has been used as a basis for action. Focused data collection is generally undervalued in terms of the direct costs of research in relation to long term benefits for conservation. It is therefore essential that partnerships prioritize data collection, and recognize the need for familiarity with changing paradigms in field methods and technologies. An example of the latter is provided by a long-term study on elephants in Mali. Ranger patrols and interviews with local people indicated the elephants no longer utilized several areas where they were previously found. Tracking of elephants using satellite GPS collars confirmed that these behavioral changes were likely permanent ranging shifts, saving effort and resources for conservation actions focused on current elephant habitats.

6.16 Impact Assessment and Monitoring

There is growing concern about the number of conservation and development interventions in Africa that are not adequately assessed for their impact, often due to their ad hoc and reactive nature. Rigorous impact assessment is essential for institutional learning about what is and is not effective in conservation and development. With the current emphasis on "adaptive management", for conservation projects to learn and adapt their strategies in response to new information, it is also essential to have some form of longitudinal data with which to compare the situation before and after an intervention. Impact assessments can also indicate unanticipated results of an intervention, whether positive or negative. This is particularly important in complex social and ecological contexts, to avoid repeating mistakes, or to reinforce actions that have had positive outcomes. Impact assessments can be used to compare alternative courses of action, by monitoring different sites or taking different actions in the same site at different times. This is valuable information for decision makers. Finally, impact assessments are essential for transparency and accountability, to show funders and local stakeholders what has been achieved with their support.

6.17 Research-Based Conservation Strategies

The challenge is to provide better-informed guidance to decision-makers about the ways in which the needs of wildlife can be reconciled with those of people and the provision of mechanisms allowing people to resolve conflicts over wildlife in ways that are not detrimental to conservation. This implies studies on the social, political, and economic factors that shape human incentives and behavior in resource use and management, and on the institutions and processes that determine how conflicts between competing interests are resolved. Typically, these are not the types of research with which wildlife biologists are familiar. Answering these research questions may require new partnerships between wildlife biologists and social scientists of various disciplines. An encouraging indicator is that some funding agencies, in particular the United States Fish and Wildlife Service, are placing increased priority on wildlife research meeting specific conservation needs.

Wildlife research must be answered with more attention to the research questions asking how to best conserve biodiversity in a cooperative strategy with the people who live with the wildlife. At present, much research is being driven by the priorities of the researchers or funders, not by the conservation needs of the areas or the questions posed by the people who live there. While basic biological information is critical, there is now a preponderance of research on factors that cause the decline of wildlife and the loss of their habitats, and on the problems faced by local people in dealing with damage from wildlife to crops and livestock. Too often, these are studied in isolation from the broader context of decisions on land use and development or from the needs of natural resource users, for example, to poor or indigenous communities who depend on natural resources for their livelihoods. The findings of this type of research are all too familiar.

6.17.1 The Role of Education

The concept of MAB (Man and the Biosphere) has been applied to many African countries, and biosphere reserves and national parks serve as ideal areas for environmental education. These are principal areas where education of the local people and the public at large can take place. This can involve the design of specific environmental education sites or, as in the case of the Malawi-Zambia Transfrontier Conservation Areas, designing curricula and educational materials for use in local schools. The education of local people through national park and game reserves programs often ties in with capacity building and training, since a large part of educating local people is to provide them with employment in the management of these areas and in research activities.

Environmental education programs are common in many African countries. Education in Uganda is being tackled through the way of "Wildlife Clubs" in many of the schools. These clubs involve children in extracurricular activities and raise the ideas and importance of wildlife conservation. Similar programs are being run in Kenya. For example, the "Young Rangers" program run by the Rhino Ark Charitable Trust is a nature and wildlife club. It is aimed at school children living in the mountain forest regions and is designed to create an understanding of the importance of the Aberdare ecosystems, thus fostering positive attitudes in the conservation of the environment.

Across Africa, formal and informal educational programs are being used to enhance public awareness and advocacy of conservation issues on a broad scale. Educational programs serve many functions for conservation, and this section explores some of the different forms of education and their impact on conservation efforts.

6.18 Environmental Education Programs

Environmental and conservation education is critical if a long-term impact on conservation is to be achieved. Few conservation projects have clearly defined how environmental education can lead to conservation action, though one activity that has been successful has been the involvement of schools through the adoption of charismatic megafauna, such as World Wildlife Fund for Nature (WWF) in Cameroon and Dian Fossey Gorilla Fund International in Rwanda, both of whom run programs with local schools. The programs have the advantage of sustainability, in that they change attitudes of future decision-makers, and have a growing constituency, as the message is taken home to families. WWF in Cameroon has just completed an environmental textbook for secondary schools, which is a great leap in integration of conservation education into national curriculum. The potential of this activity is gauged by the increase in government interest in environmental education which has resulted in an increased budget allocation for WWF to continue the work.

6.19 Capacity Building and Training

CAPRECON is careful to identify the training needs of its staff and facilitates training opportunities for many of them. In recent years, these have included participation in short courses on participatory development, conservation and development, project planning and protected area management, as well as attachments to other NGOs to learn specific skills such as community mapping or natural resources inventory. CAPRECON has also experimented with a highly innovative approach to increasing conservation training among university and college students. In 1999, it introduced a new course in environmental impact assessment for third-year students in the Faculty of Natural Resources at the University of Swaziland. This was developed and taught by CAPRECON staff in collaboration with faculty lecturers and is now an established part of the university curriculum and an ongoing relationship between CAPRECON and the university.

The aim of capacity building and training is two-fold. Partnership staff need to increase their understanding of conservation issues, and they need to improve their abilities to plan and implement conservation projects. This should enable them to be more effective in their work. Secondly, by involving a wide range of other African institutions and providing training for them, it should be possible to influence conservation outside of the immediate project areas. This might range from providing training in protected area management for government wildlife department staff, through helping other NGOs to develop their project planning skills, to raising awareness of conservation issues among universities and the academic community.

6.20 Community Outreach and Awareness Campaigns

An awareness campaign to change farming practice harmful to birdlife in the Rupununi, Guyana involved working closely with a group of indigenous communities. Farming was identified as the most important issue to address with the communities, and discussions were held with a representative from each village to decide on the best methods to achieve the desired changes in farming practice. It was agreed that education was the best approach, and two Guyanese MSc students conducted school visits to farm and non-farm households to assess bird knowledge, identify farming practices and time periods known to be harmful to birds, and to provide information and encourage change. A coloring competition was held for the younger children, and each visit ended with a village meeting during which the project was evaluated and villagers were encouraged to become actively involved. The students produced children's resources including a short illustrated story with a strong conservation message, and the campaign concluded with a second village meeting and a radio broadcast featuring a Guyanese folk musician, with discussion and live music related to birdlife and farming. Further monitoring will indicate if these activities have been successful in changing farming practice and increasing awareness about birdlife.

These campaigns intend to reach an entire community and can be addressed at various issues and using a range of media. The start of any awareness campaign would involve research to identify the target audiences and their existing knowledge/attitudes. By identifying the gaps, the campaign can be tailored to provide information where it is needed. The Carter Centre's River Blindness Control Programme in West Africa conducted an ethno-entomological study to obtain information on the population's existing knowledge and attitudes towards the blackfly in order to decide on the best strategies to raise awareness about the link between the fly and the disease, and to promote the community's active involvement in control activities.

References

Abukari H, Mwalyosi RB (2020) Local communities' perceptions about the impact of protected areas on livelihoods and community development. Glob Ecol Conser 22:e00909. https://doi.org/10.1016/j.gecco.2020.e00909

Armitage D, Mbatha P, Muhl EK, Rice W, Sowman M (2020) Governance principles for community-centered conservation in the post-2020 global biodiversity framework. Conserv Sci Pract 2(2):e160. wiley.com

Cadman R, MacDonald BH, Soomai SS (2020) Sharing victories: characteristics of collaborative strategies of environmental non-governmental organizations in Canadian marine conservation. Mar Policy 120:104129. https://doi.org/10.1016/j.marpol.2020.104129

Canton H (2021) United Nations Environment Programme—UNEP. In: The Europa directory of international organizations 2021. Routledge, pp 188–214 [HTML]

Dawson NM, Coolsaet B, Sterling EJ, Loveridge R, Gross-Camp ND, Wongbusarakum S et al (2021) The role of Indigenous peoples and local communities in effective and equitable conservation. Ecol Soc 26(3):19. https://doi.org/10.5751/ES-12625-260319

Desai BH (2020) 14. United Nations Environment Programme (UNEP). Yearbook of International Environmental Law. [HTML]

Filer C, Mahanty S, Potter L (2020) The FPIC principle meets land struggles in Cambodia, Indonesia and Papua New Guinea. Land 9(6):208. https://doi.org/10.3390/land9060208

Ivanova M (2020) Coloring the UN environmental: the catalytic role of the UN environment programme. Glob Gov26(2):307–324

Kalvelage L, Bollig M, Grawert E, Hulke C, Meyer M, Mkutu K, Diez JR (2021) Territorialising conservation: community-based approaches in Kenya and Namibia. Conserv Soc 19(4):282–293. lww.com

Koloma Y (2021) Financial inclusion and entrepreneurship willingness of youth: evidence from Mali. Afr Dev Rev 33(S1):S196–S210. https://doi.org/10.1111/1467-8268.12516

Lindsey P, Baghai M, Bigurube G, Cunliffe S, Dickman A, Fitzgerald K, Robson A (2021) Attracting investment for Africa's protected areas by creating enabling environments for collaborative management partnerships. Biol Conserv 255:108979. stevecunliffe.com

Michaelowa A, Hoch S, Weber AK, Kassaye R, Hailu T (2021) Mobilising private climate finance for sustainable energy access and climate change mitigation in Sub-Saharan Africa. Clim Policy 21(1):47–62. [HTML]

Middleton J, Cassell JA, Colthart G, Dem F, Fairhead J, Head MG, Stewart A (2020) Rationale, experience and ethical considerations underpinning integrated actions to further global goals for health and land biodiversity in Papua New Guinea. Sustain Sci 15:1653–1664. springer.com

Musavengane R, Kloppers R (2020) Social capital: an investment towards community resilience in the collaborative natural resources management of community-based tourism schemes. Tour Manag Perspect 34:100654. https://doi.org/10.1016/j.tmp.2020.100654

Nohlová B (2020) Raising the effectiveness of conservation programs: a case study of TKCP environmental and developmental activities in the YUS area, PNG. Czech Slovak J Humanit 2(1):45–62

Ofori IK, Cantah WG, Afful JrB, Hossain S (2022) Towards shared prosperity in sub-Saharan Africa: how does the effect of economic integration compare to social equity policies? Afr Dev Rev 34(1):97–113. researchgate.net

Ponte S, Noe C, Mwamfupe A (2021) Private and public authority interactions and the functional quality of sustainability governance: lessons from conservation and development initiatives in Tanzania. Regul Gov 15(3):755–772. https://doi.org/10.1111/rego.12327

Siyambango N, Enkono AN (2022) The role of local communities in resource management: a case of Okongo Community Forest and Conservancy, Namibia. J Stud Humanit Soc Sci 11(1&2):136–160. unam.edu.na

Wenborn M, Svensson MS, Katupa S, Collinson R, Nijman V (2022) Lessons on the community conservancy model for wildlife protection in Namibia. J Env Dev 31(4):375–394. sagepub.com

Zhao J, Jin T, Zhang P, Krott M, Liu J (2024) Political embeddedness in public–private partnership for nature conservation: a land trust reserve case from China. Ambio 53(2):289–302. https://doi.org/10.1007/s13280-023-01876-9

7 Wildlife-Friendly Farming

7.1 Wildlife-Friendly Farming Practices

Measures towards enhancing soil health will also help with pest problems, and crop rotation can also be a very effective method of pest prevention. In general, the less monoculture there is on a farm, the more diverse its surrounding habitats, and the more resistant it is to change, the easier it is to prevent pest problems. Integrated Pest Management fits into movement of the Entry Level Scheme (ELS), which is targeted at maintaining the high environmental value of priority and other key species and habitats. Any pesticide use should be considered a last resort and only those deemed necessary to be used for the protection of an uncropped area of qualifying land.

Integrated Pest Management (IPM) is a plan that is used to address specific pest problems, which also takes into account the impact that any method of pest management may have on the economic threshold of a pest in the long term and on other pests that are not a problem. IPM plans are constantly updated and changed in response to monitoring and research and often readdress the problem using the same method. Ideally, the idea is for every farmer to have an IPM plan, which would include pest prevention methods and emergency pest treatments and would be used as a record for the management of pest problems.

If a pest problem arises on a crop that has been managed organically, then that crop can be sold to a conventional farmer, as long as it is known what substances have been used on the crop. However, if cross-compliance is to be achieved, then this problem would have to be addressed by the farmer, who should seek to find an alternative chemical-free solution or manage the crop so that the pest is not a problem in the following years.

Reducing pesticide use is a key method in wildlife-friendly farming, as pesticides are known to have a wide range of toxic effects on birds, mammals, invertebrates, and plants. Different pesticides vary in the amount of damage they cause, but these differences are negligible when compared with the difference between pesticide use and non-use. Therefore, it is recommended that wherever possible, non-chemical alternatives to pesticides are used and to only resort to chemical methods as a last resort. If chemical pesticides are considered necessary, then it is advised that the narrowest spectrum pesticide available is used, since they are less damaging than broad-spectrum pesticides. In addition, as much care as possible should be taken to avoid spray drift.

7.1.1 Reducing Pesticide Use

Pesticides are applied to crops to eliminate potential plant damage from insects, weeds, and diseases. Unfortunately, pesticides have a greater impact

E. D. Wiafe, *Wildlife Conservation and Agriculture Development in Africa*, Sustainable Development Goals Series, https://doi.org/10.1007/978-3-032-09233-5_7

on the environment and wildlife than the benefits achieved by their use. Reducing overall pesticide usage is a central feature of wildlife-friendly farming. There are two methods of achieving this aim: the first is to change the type of pesticide used, while the second method is to utilize the alternative pest control options. Changing to softer (less toxic) pesticides often can still effectively control pests while minimizing the impact on the immediate environment. Less toxic pesticides are also less harmful to beneficial animals and insects. If only one method of pest control is utilized, it is often the case that the pest becomes resistant to that method. Using a variety of pest control options is an effective method of preventing pest resistance to any one method. A technique called "habitat manipulation" can be used to encourage natural pest control agents such as birds and spiders. This involves the creation of environments which support natural enemies of the pests. An example is leaving strips of wild plants around fields; the plants provide food and shelter for natural enemies of crop pests. This method is a long-term approach; once the correct habitats are established, they will be self-sustaining and require little maintenance.

7.1.2 Implementing Integrated Pest Management

In a recent times, bird conservationists learned IPM techniques from pest management specialists. They were encouraged by an experiment that showed how a simple change in the timing of insecticide application could have positive effects for bird populations. In the test, without reducing the amount of pesticide used, farmers left a 6-meter-wide strip of unsprayed crop at the field edge. This significantly increased the abundance, species richness, and diversity of the bird community. The insectivorous bird populations in the intervened areas increased, as birds capitalized on the higher food supply with increased nesting success. These areas became refuges for birds, which were, in addition, effective in controlling crop pests. By creating ideal conditions for birds, farmers were able to indirectly control pests and reduce the need for pesticide, essentially using the birds as free labor. The importance of this study was in showing that wildlife can be conserved and integrated into farming systems without sacrificing economic goals. An application of this model in diverse farming systems could shift the intense chemical pest management often used towards strategies that are less harmful to wildlife.

When farmers aim to reduce or eliminate synthetic pesticide use, they must be prepared for potential pest outbreaks. One way to manage pests without using toxic chemicals is through integrated pest management (IPM). IPM is a strategy that uses a variety of complementary methods: natural predators, pest-resistant crop varieties, cultural practices, and judicious use of less toxic chemicals to keep pest numbers low enough to prevent unacceptable damage or annoyance to humans and animals. These methods are implemented in a site-specific manner, monitoring the pest population and making informed management decisions based on its dynamics.

7.1.3 Enhancing Soil Health

Soil organic matter is a key indicator of soil health. It is a source of energy and nutrients for soil organisms. It improves soil structure and water holding capacity and is important in nutrient cycling. Arable soils have generally lost organic matter since the conversion of land to arable (from woodland or grassland) and the intensification of agriculture. Methods of maintaining or increasing soil organic matter, such as the use of organic manures, leaving crop residues or reduced tillage, are beneficial to wildlife. For example, farmyard manure is an important habitat for many insects such as dung beetles, which are now uncommon due to changes in farming practices. In places where it is practiced, mixed farming is beneficial to soil organic matter due to the import of organic materials from other ecosystems.

Healthy soils can more easily support a diverse community of soil organisms, from earthworms to fungi to insects, all playing essential roles in the breakdown and recycling of organic matter and in the cycling of nutrients between the soil and plants. The importance of soil health in encouraging wildlife on-farm has been increasingly recognized in recent years. Practices that enhance soil health have the potential to benefit wildlife both directly, as a provider of habitat and a living environment, and indirectly by facilitating reductions in agrochemical use and healthier plant growth.

7.1.4 Promoting Crop Rotation

A break crop refers to a crop that effectively "breaks up" a sequence of a particular type of crop. An example might be the growing of legumes, which are well known for their nitrogen-fixing abilities and are an important food source for many declining bird species. The legumes would make an excellent break crop for a sequence of cereals. Fallow has traditionally been a time when wildlife has thrived, but the intensification of agriculture has greatly reduced the extent of fallow land. Though it is not the most economically beneficial option, there are some government incentives to encourage farmers to leave land fallow, and it can have substantial benefits for wildlife. An experiment conducted in southern England compared the means of leaving arable land fallow against the natural regeneration of heathland. The fallow treatment provided nesting habitat for 85 bird species compared to 46 for the regenerating heathland. Fallow land can be further enhanced by actively managing it for wildlife.

By design, row crops are not favorable to wildlife, so one of the aims of wildlife-friendly farming is to limit the duration of the growing period of these crops (especially during the growing season) and to provide the most beneficial alternative crop and non-crop options. An example of an alternative to a detrimental row crop might be sown cereals such as guinea corn. At the other end of the spectrum, the best non-crop option might be the set-aside of arable land. Crop rotation can achieve both crop and non-crop diversification by varying the type and growing duration of different crops, and it will be beneficial to match the most suitable crop with the most suitable area of land.

Row crops are crops that are planted in rows wide enough to allow machinery or human labor to cultivate and manage the crops throughout their growth cycle. These crops are typically grown in large-scale agricultural settings and include a variety of grains, vegetables, and legumes. The organized row planting facilitates practices such as weeding, irrigation, fertilization, and harvesting.

Common Examples of Row Crops:

- **Cereals:** Maize (corn), wheat, rice, barley, sorghum
- **Legumes:** Soybeans, beans, peas, groundnuts (peanuts)
- **Vegetables:** Tomatoes, onions, cabbage
- **Cash Crops:** Cotton, tobacco, sugarcane
- **Root Crops:** Potatoes, sweet potatoes, cassava

In Africa, row cropping is widely practiced, especially for staple and cash crops. Some common African row crops include:

- **Maize (corn):** A staple food in many African countries.
- **Sorghum and Millet:** Especially in semi-arid regions.
- **Cassava and Yam:** Often planted in rows, particularly in West Africa.
- **Groundnuts (Peanuts):** Commonly grown in rows for both food and oil production.
- **Rice:** In areas with suitable water resources, such as West Africa.
- **Beans and Cowpeas:** Important sources of protein.
- **Vegetables:** Like tomatoes, onions, and cabbages in commercial farms.
- **Cash Crops:** Cotton, cocoa (though not typically in straight rows), coffee, and sugarcane.

Row crops are a dominant feature of modern farming, and it is the period when these crops are growing that wildlife is most limited in terms of available food and nesting/roosting sites. As a result, many bird species have shown large declines, often due to the intensification and increased specialization of agriculture, which usually involves the replacement of mixed farming and arable/pastoral rotations with continuous arable cropping. During the part of the rotation when row crops are grown, the weed seeds and invertebrates that many bird species depend on are in short supply, and at this critical time for breeding birds, the nesting cover is often inadequate. The removal of set-aside, which provided an important source of bird food and breeding habitat, is only likely to exacerbate this trend.

Diverse cropping systems are an essential component of wildlife-friendly farming. One of the simplest and most effective ways to diversify crops is to introduce crop rotation. This is the regular alteration of different crops in time and space on a piece of land. Typically, the rotation is made up of row crops (e.g. maize, sugar beet) and a break crop or a period of the land being fallow. The duration of the rotation cycle is very flexible; it can be annually, bi-annually, very long term, or anything in between. The benefits of crop rotation to wildlife are delivered through the spatial and temporal diversity it brings to cropping systems. In turn, this increases the variety of habitats and food sources available to farmland wildlife.

7.1.5 Implementing Precision Agriculture Techniques

Farmer interviews suggest that there is a low adoption of precision agriculture for wildlife due to a lack of confidence in new technologies and a concern about practicality and cost. To alleviate this, wildlife-friendly initiatives might concentrate on monitoring the movement of subsidies to create and maintain non-productive habitats. This might involve selecting environmental stewardship options that are aimed at specific wildlife species or ecosystems and relocating high-value farm enterprises to land of higher agricultural suitability. Data gathered for these activities can be used to assess their effect on wildlife and compare with previous levels of damage to judge success and decide if continuing investment is worthwhile.

Another important aspect of wildlife-friendly farming is predicting the likelihood of wildlife damage and varying crop protection management accordingly. This can be achieved through the use of technology to map and monitor wildlife populations in land surrounding or within the farm. The information gathered can then be used to quantify the level of wildlife damage, forecast future damage, and investigate different methods to prevent damage.

In the case of wildlife-friendly farming, it involves using precision agriculture for wildlife conservation. An information technology-based system to vary management within a field can be used to target specific wildlife at risk.

Precision agriculture is a set of site-specific management strategies that optimize farm resource use for sustaining the economic and environmental health of farm operations. It relies on new technologies like satellite imagery, information technology, and intelligent machines. This allows farmers to input data and decisions into a computer that offers a recommendation specific to the farm's requirements. Implementing precise agriculture involves a good understanding of the suitability, variability, and limitations of the above technology to the farm system.

7.2 Creating Habitat for Wildlife

The most significant impact on wildlife habitat that can be made by a farmer is to leave some of the farm in a natural state. This may involve the conversion of crop or pasture land back to native forest, or leaving areas of remnant forest or wetland uncleared. These habitats provide homes for native wildlife as well as opportunities for the control of pest animals such as sheep, cows, and goats. Where the farm contains areas of native vegetation or significant areas of regenerating vegetation, farming practices can be modified to enhance the suitability of these sites for wildlife. A very important step is to protect these areas from stock.

Creating habitat for wildlife entails providing a suitable environment for various species of mammals, birds, amphibians, reptiles and the numerous invertebrates. The aim is to encourage these animals to take up residence on the farm, and in doing so, many will seek to escape from less suitable neighboring environments to live on the more inviting farm. Habitat can be provided in each landscape feature present on the farm, ranging from the crop and pasture lands to the waterways and forested areas. Each different type of habitat will provide suitable living conditions for different types of animals.

7.2.1 Planting Native Vegetation

Native plants can also break up hardpan soil, creating an improved soil structure that allows better water infiltration and root penetration. This allows for more efficient farming in the future. Hardpan soil is often tilled repeatedly to break it up, but this creates a compaction layer which leads to increased surface water runoff. Native vegetation has the added bonus of providing farmers with a management tool for areas that are too wet to farm. These areas can be planted with native grasses and forbs to provide wildlife habitat and prevent soil erosion.

Native vegetation, once established, can take care of itself for several years without any maintenance. It also acts as a permanent cover for wildlife, providing food and habitat. Because it is perennial, native vegetation has an extensive root system that holds soil in place, preventing erosion and nutrient runoff. In contrast, non-native annual plants have shorter roots so they cannot hold soil as effectively. The root systems of native plants also allow water to infiltrate the soil more effectively, which reduces surface water runoff. This is important because nonpoint source pollution, the number one threat to water quality, occurs when surface water runoff carries excess nutrients and chemicals into streams and rivers.

A que can be taken from this: When Europeans first settled in North America, they cleared the land of its original vegetation to grow crops. As the population spread westward, almost all of the native vegetation was removed. As a result, less than one percent of the original native landscape is left in many Midwest states. Today, federal dollars are being spent to encourage farmers to set aside some of their cropland for the planting of native vegetation. There are many benefits to planting native vegetation instead of non-native plant species.

7.2.2 Establishing Hedgerows

In order to benefit wildlife, it is the structure and composition of the hedgerow that are of fundamental importance. Newly planted hedges consist of a line of shrubs, while traditional older hedges may have trees growing through them. Farmers aiming to create a hedgerow or improve an existing one for the purpose of wildlife habitat can do so in the same way as they would if they were planting a hedgerow for stock-proofing reasons. The hedges should be wide, dense, and tall. This can be achieved by planting through the hedge—allowing some of the trees to grow to full maturity—and laying, the traditional pruning of trees to ground level and laying of pleachers to form a stock-proof barrier which in due course will regenerate into a dense line of bushes and young trees. To avoid the need to trim the hedge into shape in future years, overhanging trees or those in the immediate vicinity of the hedge should be pollarded to avoid excessive shading. Where possible, hedges should be left untrimmed for a number of years as this allows the growth of lateral branches and increases the density of the hedge. A hedge is often trimmed for the first time when it has grown to the desired height.

Hedges are a familiar sight in many farming landscapes and originated from Britain, although the concept spread from there to many regions abroad. They are planted barriers, usually made up of a line of shrubs or trees, and are often protective structures that separate fields, farmsteads, or houses. Hedgerows have been planted for a number of reasons, including land enclosure, stock-proofing, and meeting the demand for timber and firewood, but any practical or aesthetic use for hedges coincides with their value as wildlife habitat. On livestock farms, the hedgerow often represents the frontier between the intensely managed pastures and meadows and the woodland and bush habitats from which stock has been excluded.

7.3 Creating Habitat for Wildlife

7.3.1 Creating Wetlands and Ponds

Compared with other habitat types, wetlands provide disproportionately high levels of biological diversity. This is a powerful statement. Many studies have proven the fact that wetlands are exceedingly important for a vast array of wildlife species. Unfortunately, the rate of wetland destruction has left many species without suitable habitat. In many agricultural areas, it is possible to enhance or create wetlands for wildlife. Certain land that is poorly drained or prone to flooding is not the best for agriculture. By restoring wetlands and ponds, the farmer can actually improve the agricultural value of the land in addition to providing wildlife habitat. Wetlands purify water, recharge aquifers, and provide flood control. Many government programs exist to aid in wetland restoration. Contact the Natural Resource Conservation Service, the U.S. Fish and Wildlife Service, or the local Department of Natural Resources office for more information about wetland restoration programs.

7.3.2 Providing Food Sources and Water

To attract and support wildlife, there must be an ample supply of food. This includes food in the form of plants, insects, and other invertebrates, which make up the diet of the majority of terrestrial wildlife. Some large mammals and most birds also eat various seeds. Many of these plant and insect food sources are naturally occurring and simply need to be preserved. Others can be created through habitat enhancement. A simple way to add food sources to a farm is to plant native trees, shrubs, and herbaceous vegetation. Another way is to set aside a portion of the farm as a wildlife food plot or field. This can be done by leaving a portion of the farm fallow for a number of years and allowing natural vegetation to grow, or by planting seed from native grasses, forbs, or shrubs. These areas can be enhanced by allowing a limited amount of livestock grazing, or by mowing, burning, or using other vegetation management techniques to promote new and succulent growth in the plants. This will create a mosaic of different vegetation types and successional stages, and will attract different types of wildlife. For example, weaver birds and quail feed on the seeds of various grasses and forbs. Duikers eat browse from many types of woody plants. Providing food for wildlife can also be accomplished through livestock management. By altering the timing and intensity of livestock grazing on certain pastures or rangelands, it is possible to increase the food supply for many types of wildlife. In addition to increasing the food supplies on the farm for wildlife, it is also important to decrease or eliminate any supplemental feeding of wildlife, especially in the case of game species. Feeding wildlife can artificially increase their numbers or keep them in an area where they would not otherwise be able to find adequate food, resulting in damage to the land and vegetation.

7.4 Case Studies in Agroecology

7.4.1 Case Study: Agroecology and Biodiversity—The Push-Pull Technology in East Africa

Agricultural intensification has often been at odds with wildlife conservation in Africa, leading to biodiversity loss, habitat degradation, and soil depletion. However, agroecological innovations like **Push-Pull Technology (PPT)** present a sustainable alternative by balancing food production with ecosystem health. This case study explores how PPT—a nature-based pest control and soil management system—enhances biodiversity, conserves beneficial insects, and reduces the environmental footprint of farming in East Africa.

7.5 Background and Ecological Challenges

East African smallholder farmers depend on maize and sorghum for food security, but their productivity is severely threatened by:

Stemborer moths (*Busseola fusca, Chilo partellus*)—Major cereal pests causing extensive crop losses.

Striga weed (*Striga hermonthica*)—A parasitic plant that suppresses cereal yields and spreads rapidly.

Soil degradation—Loss of soil nutrients due to unsustainable farming practices.

Chemical pesticide overuse—Leading to declines in pollinators and beneficial insects.

To address these challenges, the International Centre of Insect Physiology and Ecology (icipe), in collaboration with Rothamsted Research, developed Push-Pull Technology as an agroecological solution that integrates pest management, soil restoration, and biodiversity conservation.

PPT employs natural plant interactions to create a self-sustaining pest control and soil fertility system:

Push Effect: Intercropping cereals with Desmodium, a leguminous plant that repels stemborers, suppresses Striga, and improves soil fertility by fixing nitrogen.

Pull Effect: Surrounding fields with Napier grass (*Pennisetum purpureum*), which attracts stemborers away from the main crop and acts as a trap, reducing pest populations.

This system reduces the need for synthetic pesticides and fertilizers, promoting a more balanced ecosystem that supports insect biodiversity and improves soil health.

7.6 Conservation and Agricultural Development Benefits

Restoration of Beneficial Insects and Pollinators: Unlike chemical pesticides that harm pollinators, PPT fosters a thriving environment for bees, butterflies, and predatory insects that contribute to natural pest control.

Wildlife Habitat Creation: Napier grass and Desmodium provide cover for birds, small mammals, and beneficial arthropods, enhancing farmland biodiversity.

Reduced Land Degradation and Soil Fertility Improvement: Leguminous Desmodium replenishes soil nitrogen, reducing the expansion of agriculture into wildlife habitats caused by soil exhaustion.

Food Security and Climate Resilience: Farmers using PPT report two to three times higher maize yields, ensuring food security while mitigating the effects of drought and climate change.

Minimized Chemical Runoff and Water Contamination: By replacing synthetic pesticides with natural pest control, PPT reduces toxic runoff into rivers and wetlands, protecting aquatic biodiversity.

7.7 Challenges and Future Directions

Adoption Barriers: Access to Desmodium and Napier seeds, as well as farmer awareness, limit large-scale implementation.

Policy and Institutional Support: Scaling up PPT requires investment in farmer training and government incentives for agroecological practices.

Integration with Wildlife Corridors: Further research is needed to explore how PPT can be integrated with conservation corridors to enhance coexistence between agriculture and wildlife.

Push-Pull Technology exemplifies how agroecology can bridge the gap between wildlife conservation and agricultural development in Africa. By harnessing ecological interactions instead of chemical inputs, PPT promotes biodiversity, improves soil health, and enhances farmer livelihoods. As Africa navigates the challenges of food security and environmental sustainability, such innovations will be crucial in shaping a future where agriculture and conservation coexist (Khan et al. 2021; Pickett et al. 2014).

7.7.1 Organic Farming and Biodiversity Conservation

Organic farming began as an outcome of the environmental conservation movement in the 1970s. It was a direct action to counter the negative externalities of the Green Revolution, which many small-scale farmers and environmentalists saw as a threat to biological diversity. The elimination of chemical pesticides and fertilizers and the simplification of farm management, which are key components of organic agriculture, may directly affect farmland wildlife in a range of positive methods. We consider how present and in-depth organic agriculture affects wildlife habitat, species diversity, and farmland ecosystem dynamics. Studies of the impact of land use change to organic farming on wildlife are few, and much of the concepts about the environmental effects of organic farming are largely theoretical. This reflects the global boom in organic agriculture and the significance of moving farming systems in this direction, but needs to be addressed if we are to weigh up the pros and cons of converting land to organic agriculture in terms of conservation. Organic farming at present is not an entirely homogenous entity, and a variety of farming systems exist under the organic banner, e.g., low-external input and high biodiversity farms (LEIHB), biodynamic farming, pragmatic organic. To some extent, practices combining the conservation of wildlife and habitat, with the production of food and fiber, may be relevant to all types of organic farming. Yet farming in the interest of wildlife is more likely to occur on specialized wildlife farms or on land managed primarily for conservation. On intensively farmed land where agriculture is the sole priority, organic farming is most likely to benefit wildlife through a reduction in inputs and the abandonment of land. This presents a positive option to land sparing for biodiversity if it can be linked to reversion or direct conservation of some land to a wilder state.

Africa's rich biodiversity and agricultural landscapes face growing threats from deforestation, land degradation, and chemical-intensive farming. Organic farming, which avoids synthetic fertilizers and pesticides, offers a sustainable alternative that aligns with biodiversity conservation while supporting food security and rural livelihood

Organic farming promotes soil fertility through composting, cover cropping, and crop rotation. These practices improve soil organic matter, water retention, and microbial diversity, making farmlands more resilient to drought and climate change. Healthy soils support a wider range of plant and insect species, enhancing ecosystem stability.

Africa's agricultural productivity heavily depends on pollinators like bees, butterflies, and birds. The widespread use of pesticides has led to their decline. Organic farming eliminates harmful agrochemicals, allowing pollinator populations to recover, which in turn boosts crop yields and biodiversity.

Many African farmers practicing organic agriculture cultivate traditional crops like millet, sorghum, and indigenous vegetables. These crops are naturally resistant to pests and diseases, reducing dependency on chemical inputs while preserving genetic diversity crucial for climate adaptation.

Organic farms often incorporate agroforestry and buffer zones that provide refuge for birds, small mammals, and reptiles. Unlike monocultures, organic landscapes create mosaic habitats, supporting both agricultural productivity and conservation efforts.

With rising global demand for organic products, African farmers have access to premium markets in Europe, North America, and Asia. Countries like Uganda, Ethiopia, and Tunisia have growing organic sectors, benefiting smallholder farmers through higher export prices and fair-trade certifications.

While organic farming has significant benefits, challenges such as low adoption rates, lack of technical support, and limited access to organic inputs hinder its widespread implementation. Governments and research institutions must invest in policy incentives, farmer training, and organic certification systems to unlock its full potential.

Organic farming presents a unique opportunity to harmonize food production with biodiversity conservation in Africa. By promoting soil health, protecting pollinators, and preserving indigenous crops, it contributes to sustainable agriculture while safeguarding Africa's ecological heritage. As awareness and market demand grow, organic farming can serve as a key pillar for Africa's agricultural future.

7.7.2 Agroforestry Systems and Wildlife Habitat

Tropical agroforestry systems are often portrayed as the best of all worlds for wildlife, domestic animals, and people. The distinguishing characteristic of agroforestry is that it incorporates both trees and herbaceous plants or animals in a complex of spatial and temporal arrangements. A simplification of this definition could be the intentional combination of woody perennials and agriculture or livestock on the same unit of land. This wide range of practices can be found in Africa and other regions of the world. But the greatest diversity of agroforestry systems, and the most complex and least understood, are found in the moist tropics. This is where most of the world's biodiversity is also found. But the conservation of biodiversity in agroforestry systems is paradoxically one of the least studied areas of agroforestry, though this situation is rapidly changing. Biodiversity itself is a very broad topic; here the focus should be on tropical forest and its associated flora and fauna. As Africa grapples with the twin challenges of **agricultural expansion and biodiversity loss**, agroforestry has emerged as a vital strategy for integrating conservation into farming landscapes. Unlike conventional monocultures, agroforestry systems mimic natural ecosystems by incorporating trees, crops, and livestock, thereby providing habitats for wildlife, enhancing soil fertility, and improving ecosystem resilience. This chapter synthesizes current knowledge on how different types of agroforestry impact African biodiversity and explores ways to modify these systems to better conserve wildlife while sustaining agricultural productivity.

Agrisilviculture (Trees and Crops)—where trees and crops are grown together—is widespread in cocoa and coffee farms in Ghana, Ethiopia, and Uganda. These shaded systems provide microhabitats for birds, insects, and small mammals, mitigating the habitat destruction often associated with open-field farming. The positive effects are: (i) maintenance of tree cover, reducing habitat fragmentation; (ii) supporting pollinators like bees and butterflies, essential for crop yields; and (iii) provision of food and shelter for seed-dispersing species, aiding forest regeneration. The challenges of the system are: (i) commercial pressure leads to monoculture tree plantations (e.g., Eucalyptus, Pine), which support lower biodiversity and (ii) deforestation for farmland expansion reduces old-growth trees critical for nesting birds and primates.

Silvopastoral Systems (Trees and Livestock)—where trees are integrated into livestock grazing lands—are common in savanna and rangeland ecosystems of East and Southern Africa. They maintain vegetation cover, reducing land degradation while providing shade and forage for both livestock and wildlife. The positive effects are: (i) promotes diverse grasses and shrubs that support pollinators and herbivores; (ii) maintains habitat for insectivorous birds and small carnivores that regulate pest populations (iii) Enhances carbon sequestration, mitigating climate change impacts. The main challenges include (i) overgrazing can lead to land degradation and tree loss, reducing habitat quality; (ii) conversion of rangelands into industrial-scale livestock farms disrupts traditional grazing corridors for migratory species like antelopes.

Homegardens and Multi-Strata Agroforests: Smallholder farmers in West and Central Africa often maintain homegardens, where fruit trees, medicinal plants, and vegetables are cultivated together. These multi-layered agroforests replicate natural forests, making them highly beneficial for biodiversity conservation. The positive effects are as follows: (i) provides refuge for birds, pollinators, and small mammals in heavily farmed landscapes. (ii) supports indigenous crop species, preserving agricultural genetic diversity; (iii) reduces the need for chemical pesticides, protecting soil microbiota and freshwater systems.

The main challenges are (i) growing commercialization of agriculture leads to the replacement of diverse tree species with monoculture cash crops (e.g., oil palm, rubber) and (ii) the introduction of non-native species can displace native flora and fauna.

Riparian Buffer Agroforestry—planting trees along riverbanks and wetlands—is crucial for protecting Africa's freshwater biodiversity. Found in Uganda's Nile Basin and South Africa's wetland zones, these systems prevent soil erosion, filter agricultural runoff, and provide wildlife corridors. The positive effects are as follow: (i) Supports aquatic biodiversity by reducing sedimentation in rivers. (ii) Maintains nesting sites for wetland birds and amphibians; (iii) Acts as a buffer against flooding and climate variability. The key challenges of the system are: (i) encroachment for farmland expansion reduces wetland connectivity and Agrochemical pollution from nearby farms can degrade water quality.

7.8 Modifying Agroforestry Systems for Enhanced Biodiversity Conservation

To maximize biodiversity conservation, agroforestry systems must be adapted to support both wildlife and sustainable agricultural production. Key modifications include:

Promoting Native Tree Species Over Exotic Varieties: Replacing Eucalyptus and Pine plantations with native species like Baobab (*Adansonia digitata*), Iroko (*Milicia excelsa*), and Acacia enhances habitat quality and ecological functions.

Strengthening Wildlife Corridors: Agroforestry landscapes should be designed to maintain connectivity between natural habitats, allowing wildlife movement across agricultural zones.

Incorporating tree-lined hedgerows and buffer zones improves habitat structure for birds and small mammals.

7.9 Encouraging Multi-Species Agroforestry Over Monoculture Plantations

Diversified agroforests—with a mix of fruit, nut, timber, and nitrogen-fixing trees—support a broader range of pollinators, seed dispersers, and predators. Farmers can integrate **medicinal plants and indigenous crops**, reducing dependence on external inputs while boosting biodiversity.

Implementing Incentives for Biodiversity-Friendly Agroforestry: Governments and conservation organizations should introduce Payments for Ecosystem Services (PES) to reward farmers who adopt biodiversity-friendly agroforestry practices; and/or Certification schemes (e.g., Rainforest Alliance, Organic Agriculture) can provide economic incentives for sustainable agroforestry.

Agroforestry presents a viable pathway for balancing wildlife conservation with agricultural development in Africa. By maintaining tree cover, enhancing soil fertility, and providing habitats for diverse species, it offers an ecological alternative to conventional farming. However, its success in biodiversity conservation depends on modifying existing systems to prioritize native tree species, improve landscape connectivity, and integrate diverse crops. With proper policy support and farmer incentives, agroforestry can play a crucial role in securing Africa's agricultural future while safeguarding its rich biodiversity (Garrity 2012; Perfecto and Vandermeer 2010; Scherr and McNeely 2008).

7.9.1 Conservation Agriculture and Soil Conservation

Due to the practices used in Conservation Agriculture (CA), there is no doubt that it benefits farmland wildlife. As shown by a review of sustainable agriculture in developing countries, it was noted that direct-seeded, minimum tillage systems used in CA work contribute to the recovery of arthropods (by creating ideal habitat), and an increase in their abundance is beneficial in terms of bird populations. Conservation agriculture (CA) is an approach that aims to create a more sustainable and profitable production system by using fewer inputs, optimizing the use of on-farm resources, and minimizing the adverse effects on the environment. It is based on three linked principles: minimal soil disturbance, permanent organic soil cover, and diversification of crop species grown in sequences and/or associations. It makes use of the latest technologies, which combine conservation and production to change farmers' practices, and it is an effective way to increase crop productivity without harming wildlife, which is directly affected by intensive agricultural practices. Soil is the foundation of both agriculture and natural ecosystems, yet it is increasingly threatened by deforestation, overgrazing, and unsustainable farming practices across Africa. Conservation agriculture (CA) has emerged as a key strategy to enhance soil conservation while ensuring agricultural productivity. By minimizing soil disturbance, maintaining soil cover, and integrating crop diversification, conservation agriculture supports both wildlife conservation and sustainable agricultural development.

Conservation agriculture is based on three core principles: (i) Minimum Soil Disturbance (No-Till Farming) which reduces erosion by maintaining soil structure; enhances soil microbial diversity, which is essential for nutrient cycling and limits carbon loss, contributing to climate change mitigation. (ii) Permanent Soil Cover (Mulching and Cover Crops) that reduces water evaporation, improving soil moisture retention; prevents soil erosion by minimizing the impact of raindrops and wind and provides habitat for soil organisms and beneficial insects. (iii) Crop Diversification (Intercropping and Crop Rotation) that improves soil fertility by fixing nitrogen and replenishing nutrients; disrupts pest cycles, reducing the need for chemical pesticides and supports agrobiodiversity, creating a more resilient farming system.

Soil erosion is a major threat to agricultural productivity and ecosystem stability in Africa. Conservation agriculture mitigates this by:

- Reducing topsoil loss through permanent vegetation cover.
- Increasing organic matter, which improves soil structure and water infiltration.
- Protecting riverbanks and watersheds, reducing sediment runoff into aquatic ecosystems.

Conservation agriculture improves soil organic matter, leading to higher nutrient availability for crops. No-till farming stores carbon in the soil, reducing greenhouse gas emissions. Cover crops prevent nutrient leaching, ensuring long-term soil fertility.

Conservation agriculture further creates microhabitats for wildlife and promotes ecological balance:

- Birds and small mammals benefit from undisturbed crop residues and hedgerows.
- Soil-dwelling organisms like earthworms and fungi flourish in no-till systems, improving soil aeration.
- Reducing pesticide use supports pollinators and natural pest predators.

Despite its benefits, conservation agriculture faces challenges in Africa, including: (i) limited farmer awareness and technical knowledge (ii) short-term yield concerns, as soil health improvement takes time and (iii) discouraging long-term investments in soil conservation.

Conservation agriculture offers a powerful solution for balancing food production with wildlife conservation and soil health in Africa. By protecting soil, enhancing biodiversity, and promoting ecosystem resilience, CA aligns with sustainable agricultural development and long-term environmental conservation. Its widespread adoption will be critical for securing Africa's agricultural future while preserving its rich biodiversity.

7.9.2 Sustainable Livestock Management and Wildlife Coexistence

Farmers often view wildlife as competitors for forage and as pests of crops and livestock. Because of these economic and management conflicts, many wildlife species have declined from agricultural ecosystems and some have been eliminated. Wildlife damage to crops and livestock can be a significant problem in some situations; however, it is often overestimated and misunderstood (Fig. 7.1). Much of the perceived wildlife damage is, in fact,

Fig. 7.1 Elephant foraging in African savanna wildlife corridor (Photo by E.A. Tetteh)

caused by mismanagement of the farming system. In the long term, sustainable farming systems can support both viable farm enterprises and a diverse wildlife community. Key to achieving this is the development of a more informed understanding of the interactions between farming enterprises and wildlife, coupled with systems of management that are sympathetic to the needs of wildlife.

Livestock farming is a key driver of food security and rural livelihoods in Africa, supporting millions of pastoralists and smallholder farmers. However, unsustainable grazing practices, habitat encroachment, and conflicts with wildlife have led to land degradation, biodiversity loss, and competition for resources. Sustainable livestock management (SLM) offers a solution by promoting ecological balance, rangeland restoration, and coexistence between domestic animals and wildlife.

Sustainable Livestock Management Practices: This includes Rotational and Holistic Grazing and it (i) prevents overgrazing, allowing vegetation recovery and soil regeneration (ii) enhances grassland biodiversity, supporting both livestock and wild herbivores and (iii) reduces soil erosion and desertification, preserving ecosystems (Fig. 7.2).

Silvopastoral Systems (Integrating Trees with Livestock Farming): this provides shade and forage while maintaining tree cover and it is essential to support pollinators, birds, and soil organisms, improving biodiversity; and enhances carbon sequestration, mitigating climate change.

Predator-Friendly Livestock Management and this involves the use of livestock guardian animals (e.g., dogs, donkeys) to reduce predation. It helps in night enclosures and non-lethal deterrents to protect livestock from predators and provides compensation and insurance schemes to reduce human–wildlife conflict.

7.10 Wildlife and Livestock Coexistence

Restoring Wildlife Corridors this involves designing land-use plans that allow migration of wild species and establishing buffer zones between conservation areas and grazing lands.

Fig. 7.2 Movement of cattle from grazing field to kraal (Photo by E.D. Wiafe)

Community-Based Conservation and Pastoralism. This involves eco-tourism initiatives that benefit pastoral communities while protecting wildlife and provides incentives for sustainable grazing and habitat conservation.

Disease Control and Biodiversity Protection. This involves applying veterinary programs to prevent disease transmission between livestock and wildlife and restricting livestock grazing in protected areas to minimize habitat destruction.

Sustainable livestock management is essential for harmonizing food production with wildlife conservation in Africa. By adopting ecological grazing methods, protecting habitats, and promoting coexistence strategies, Africa can ensure that livestock farming supports both people and biodiversity.

7.10.1 Permaculture Principles in Farming for Wildlife

What is permaculture? The term "permaculture" was first coined by Australians Bill Mollison and David Holmgren in 1978. Permaculture, as a word, is a contraction of two words: "permanent agriculture". Permaculture is concerned with sustainable land use and design. It is a set of design principles centered on the direct utilization of, or mimicry of, patterns and features observed in natural ecosystems. This includes various agricultural methods, and no-till farming has caught a lot of interest recently. This method originated in the 1930s when the Australian government was looking at new methods to control rabbit populations. Rabbit holes were plugging up machinery used by large-scale farms, and a method called the Rome plow was being used at the time to dig out the burrows. This led to a researcher by the name of Edward Faulkner looking at ways to kill the rabbits without bringing up more soil and the development of a tool called the "no. 1 rabbit destroyer". This was a shaped explosive with a detonator and a safety fuse that killed the rabbit in its burrow. While this sounds horrible and there was mass ecological damage done to many fields, the point is that no-till agriculture originally began as a way to preserve ecosystems. With growing environmental awareness, the no-till agriculture movement has been re-evaluated in terms of lessening soil erosion and preventing damage to fragile ecosystems caused by traditional methods of agriculture. Permaculture, rooted in the philosophy of working with nature rather than against it, is increasingly being adopted in Africa to create agricultural landscapes that support both food production and wildlife conservation. Unlike conventional farming, which often leads to habitat destruction and biodiversity loss, permaculture integrates diverse plant and animal species, mimicking natural ecosystems to ensure long-term ecological balance. This approach provides a sustainable way to enhance soil fertility, conserve water, and create wildlife-friendly agricultural systems that benefit both farmers and the environment.

In many parts of Africa, smallholder farmers are designing their farms to function as miniature ecosystems, where crops, trees, and animals interact harmoniously. By cultivating a variety of plant species, farmers create layered landscapes that resemble natural forests, offering food and shelter to native wildlife. Hedgerows, windbreaks, and agroforestry systems not only provide microhabitats for birds, pollinators, and beneficial insects but also contribute to soil stability and moisture retention. The strategic placement of ponds and wetlands within farmlands attracts amphibians and aquatic species while regulating water cycles and enhancing biodiversity.

Livestock integration in permaculture-based farming follows nature's rhythms, ensuring that grazing patterns do not degrade the land but rather contribute to soil regeneration. Free-range poultry and rotational grazing systems allow herbivores to naturally fertilize the land, reducing dependence on synthetic inputs. Farmers who embrace these practices often observe an increase in predator–prey interactions, where natural pest controllers such as owls, hawks, and small carnivores thrive, reducing the need for chemical pesticides. This delicate balance fosters a farming landscape where agricultural productivity and wildlife conservation are not opposing forces but complementary elements of a resilient system.

Communities practicing permaculture often engage in seed saving, organic composting, and indigenous knowledge application to further sustain their agricultural ecosystems. By cultivating native plants and maintaining wild zones within farms, they ensure that their land remains a haven for biodiversity. In regions where large mammals and migratory species face habitat fragmentation, permaculture-based farms serve as ecological corridors, linking protected areas and allowing wildlife to move across human-dominated landscapes without conflict.

As climate change intensifies and agricultural pressures mount, permaculture presents a promising solution for Africa's farming future. By designing regenerative landscapes that restore natural functions, farmers not only secure their food sources but also contribute to the preservation of Africa's rich wildlife heritage. The success of these approaches lies in their adaptability, proving that agriculture and conservation can thrive together through mindful land management and a deep respect for nature's interconnected systems.

The UN reports state that as the global food and financial crises continue to unfold, smallholders who produce both food and cash crops can make or break real food security and economic security. Real food security is defined as access to enough healthy and culturally appropriate food produced by local farmers to sustain an active, healthy life for everyone. This report, via a permaculture lens, will show the benefits of permaculture and look at several methods that smallholders are using within these climates to produce food and cash crops while focusing on wildlife conservation. Farmers who work in the permaculture/agroecology field are on the frontlines of true food security as our truly sustainable methods are good for the planet, provide food security for the farmer, and have the ability to scale across a wide range of ecosystems worldwide.

References

Garrity DP (2012) Agroforestry and ecosystem services in Africa. J Sustain Agriculture 36(2):134–152

Khan MA, Ali S, Khan MA, Ali A (2021) Adoption of climate-smart agriculture practices: evidence from smallholder farmers in Pakistan. J Clean Prod 298:126804. https://doi.org/10.1016/j.jclepro.2021.126804

Perfecto I, Vandermeer J (2010) Agroecology and the conservation of biodiversity. Conserv Biol 24(3):846–850

Pickett JA et al (2014) Striga control and climate resilience through a cereal–legume intercrop system. Nat Clim Change 4(2):172–177

Scherr SJ, McNeely JA (2008) Biodiversity conservation and agricultural sustainability: towards a new paradigm of ecoagriculture landscapes. Philos Trans R Soc B 363(1491):477–494

8 Policy Frameworks Promoting Wildlife Conservation and Food Security in Africa

Balancing wildlife conservation and food security in Africa requires strong policy frameworks that integrate biodiversity protection with sustainable agricultural practices. Many African nations have enacted policies to safeguard wildlife while ensuring food production remains resilient amid environmental and socioeconomic challenges. However, the effectiveness of these policies depends on their implementation, enforcement, and adaptability to local contexts. This discussion explores key policy frameworks that govern wildlife conservation and food security in Africa and assesses their implementation and impact.

8.1 Wildlife Conservation Policies in Africa

Wildlife conservation policies in Africa are largely guided by national legislation, regional agreements, and international conventions. Key frameworks include the African Convention on the Conservation of Nature and Natural Resources (Algiers Convention), the Convention on Biological Diversity (CBD), and the Convention on International Trade in Endangered Species of Wild Fauna and Flora (CITES). Many African countries have established national parks, wildlife reserves, and community conservation areas under these frameworks to protect endangered species and critical ecosystems.

Governments have also developed national biodiversity strategies to mitigate habitat loss, regulate hunting and poaching, and encourage community-based conservation initiatives. Policies promoting ecotourism and sustainable land use have created economic incentives for wildlife conservation, while transboundary agreements, such as the Kavango-Zambezi Transfrontier Conservation Area (KAZA TFCA), facilitate regional cooperation in protecting migratory species.

Despite these frameworks, challenges persist. Illegal wildlife trade, weak law enforcement, and human–wildlife conflicts continue to undermine conservation efforts. Additionally, conservation policies sometimes prioritize wildlife protection at the expense of local communities, leading to resistance and unsustainable land-use practices.

8.2 Food Security Policies in Africa

Ensuring food security in Africa requires policies that promote sustainable agriculture, improve rural livelihoods, and build climate resilience. The Comprehensive Africa Agriculture Development Programme (CAADP), launched by the African Union (AU), provides a continental framework for increasing agricultural productivity while ensuring environmental sustainability. National policies, such as Ghana's Planting for Food and Jobs program and Kenya's Agricultural Sector

E. D. Wiafe, *Wildlife Conservation and Agriculture Development in Africa*, Sustainable Development Goals Series, https://doi.org/10.1007/978-3-032-09233-5_8

Transformation and Growth Strategy (ASTGS), aim to enhance food production through subsidies, improved technology, and market access.

Many policies also integrate climate-smart agriculture (CSA) and sustainable land management to prevent soil degradation and desertification. Efforts to promote agroforestry, conservation agriculture, and organic farming align with wildlife conservation by reducing deforestation and chemical pollution. However, food security policies often face hurdles such as inadequate infrastructure, political instability, and climate change-induced disruptions.

8.3 Efficacy of Policy Implementation

8.3.1 Strengths of Policy Implementation

Effective policy implementation translates policy intentions into tangible outcomes. One major strength is clarity of direction, where well-defined goals, roles, and procedures guide institutions and stakeholders, reducing ambiguity and improving coordination. Strong implementation frameworks also enhance institutional accountability, as responsibilities are clearly assigned and performance can be monitored against set targets. Another key strength is improved service delivery and impact, since policies that are effectively implemented lead to real improvements in livelihoods, environmental management, economic growth, or social welfare. Good implementation often encourages stakeholder participation, fostering ownership, compliance, and legitimacy—especially when communities, private sector actors, and civil society are actively engaged. Policy implementation further strengthens learning and adaptability. Through monitoring, evaluation, and feedback mechanisms, implementers can identify gaps and adjust strategies to changing conditions. This adaptive capacity is critical in complex sectors such as climate change, biodiversity conservation, and development planning. Finally, effective implementation enhances credibility and trust in governance systems. When policies are implemented consistently and fairly, public confidence in institutions increases, which in turn supports long-term policy sustainability and the successful introduction of future reforms.

8.4 Challenges in Policy Implementation

Weak Law Enforcement: Wildlife protection laws are often underfunded and poorly enforced, leading to continued poaching and habitat destruction.

Land Use Conflicts: Policies sometimes create tensions between conservation and agriculture, particularly where protected areas restrict land access for local farmers and pastoralists.

Climate Change Impact: Despite policies promoting resilience, droughts, floods, and extreme weather events continue to threaten food security and conservation efforts.

Policy Fragmentation: In many countries, conservation and agricultural policies are implemented by separate government agencies, leading to conflicting land-use priorities.

8.5 The Way Forward

For wildlife conservation and food security policies to be more effective, African governments must pay attention to the following:

1. Strengthen law enforcement and governance structures to combat illegal activities threatening wildlife.
2. Promote integrated land-use planning that balances conservation with sustainable farming and livestock management.
3. Invest in agroecological approaches that support biodiversity while ensuring food production remains sustainable.
4. Enhance community participation in policy-making, ensuring local knowledge and livelihoods are considered in conservation efforts.
5. Foster international partnerships to provide funding, technical expertise, and capacity-building for long-term sustainability.

Africa's policy frameworks for wildlife conservation and food security have laid an important foundation for sustainable development. However, their effectiveness depends on robust implementation, local engagement, and cross-sector collaboration. By refining policies to address socio-economic realities, strengthening enforcement mechanisms, and integrating conservation with sustainable food production, Africa can achieve a future where both people and wildlife thrive together (African Union 2003; Convention on Biological Diversity (CBD) 1992; CITES 1973; Food and Agriculture Organization (FAO) 2017; Food and Agriculture Organization (FAO) 2021; International Union for Conservation of Nature (IUCN) 2020; Kavango-Zambezi Transfrontier Conservation Area (KAZA TFCA) 2020; United Nations Environment Programme (UNEP) 2022; West and Naughton-Treves 2019).

References

African Union (2003) Comprehensive Africa Agriculture Development Programme (CAADP). Retrieved from https://au.int

Convention on Biological Diversity (CBD) (1992) Convention on Biological Diversity. United Nations. Retrieved from https://www.cbd.int

Convention on International Trade in Endangered Species of Wild Fauna and Flora (CITES) (1973) CITES Treaty. Retrieved from https://www.cites.org

Food and Agriculture Organization (FAO) (2021) Agroecology in Africa: Integrating Food Security and Biodiversity Conservation. FAO, Rome

Food and Agriculture Organization (FAO) (2017) The Future of Food and Agriculture – Trends and Challenges. FAO, Rome

International Union for Conservation of Nature (IUCN) (2020) Community-Based Natural Resource Management in Africa: A Review of Best Practices. IUCN, Gland

Kavango-Zambezi Transfrontier Conservation Area (KAZA TFCA) (2020) KAZA TFCA Strategy and Implementation Plan. Retrieved from https://www.kavangozambezi.org

United Nations Environment Programme (UNEP) (2022) Wildlife Conservation and Sustainable Agriculture in Africa. UNEP, Nairobi

West PC, Naughton-Treves L (2019) Protected areas and human livelihoods in Africa: a review of social and environmental impacts. Conserv Biol 33(4): 812–824

MIX
Papier aus verantwortungsvollen Quellen
Paper from responsible sources
FSC® C105338

If you have any concerns about our products,
you can contact us on
ProductSafety@springernature.com

In case Publisher is established outside the EU,
the EU authorized representative is:
Springer Nature Customer Service Center GmbH
Europaplatz 3, 69115 Heidelberg, Germany

Printed by Libri Plureos GmbH
in Hamburg, Germany